World Compass Academy
2490 South Perry Street
Castle Rock, CO 80104

A Journey Through History

THE STORY OF EXPLORATION AND DISCOVERY

Simon Adams

New York

This edition published in 2011 by:

The Rosen Publishing Group, Inc.
29 East 21st Street
New York, NY 10010

Additional end matter copyright © 2011 by The Rosen Publishing Group, Inc.

All rights reserved. No part of this book may be reproduced in any form without permission in writing from the publisher, except by a reviewer.

Library of Congress Cataloging-in-Publication Data

Adams, Simon.
  The story of exploration and discovery / Simon Adams.
      p. cm. — (A journey through history)
  Includes bibliographical references and index.
  ISBN 978-1-4488-0622-5 (library binding)
  1. Discoveries in geography. I. Title.
  G82.A32 2011
  910.9—dc22
                                              2010006104

Manufactured in the United States of America

CPSIA Compliance Information: Batch #S10YA: For further information, contact Rosen Publishing, New York, New York, at 1-800-237-9932.

Copyright in design, text and images © 2000, 2008 Anness Publishing Limited, U.K.
Previously published as *Exploring History: Exploration and Discovery*.

Picture Credits
The publishers would like to thank the following
artists who have contributed to this book:
Julian Baker (Baker Illustrations); Mark Bergin; Vanessa Card;
James Field (SGA); Chris Forsey; Terry Gabbey; Ron Hayward (Hayward Art Group);
Sally Holmes; Richard Hook (Linden Artists); John James (Temple Rogers); Aziz Khan;
Shane Marsh (Linden Artists); Martin Sanders; Peter Sarson; Rob Sheffield; Guy Smith
(Mainline Design); Roger Stewart; Ken Stott; Mike Taylor (SGA); Mike White (Temple
Rogers); John Woodcock.
Panel Maps: Steve Sweet (SGA).
Main Maps: Roger Stewart.

The publishers wish to thank the following
for supplying photographs for this book:
Page 38 (C/R) A. Zvoznikov/Hutchison Library; 40 (B/R)
R. Francis/Hutchison Library; 42 (T/R) N.Haslam/Hutchison Library; 42 (B/L) E.
Parker/ Hutchison Library; 42 (B/R) Hutchison Library; 44 (M/R) Hutchison Library;
46 (T/L) The Stock Market; 46 (B/R) A. Singer/ Hutchison Library; 48 (M) C. Pasini/
Hutchison Library; 49 (B/L) N. Smith/Hutchison Library; 54 (B/L) Corbis; 55 (B)
Popperfoto; 56 (B/R) Popperfoto; 58 (B/L) Popperfoto; 58 (T/R) Popperfoto/ Reuter.

All other pictures from Dover Publications and Miles Kelly archives.

Contents

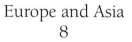

Introduction
4

Egyptians, Phoenicians and Greeks
6

Europe and Asia
8

The Invasion of Europe
10

The Vikings
12

The Polynesians
14

Crossing the Deserts
16

The Chinese Empire
18

Travelers in Europe
20

The Portuguese
22

Christopher Columbus
24

Conquering the
New World
26

Round the World
28

Into Canada
30

The Northwest
Passage
32

The Northeast
Passage
34

Exploring Asia
36

Advancing into
America
38

Across the Pacific
40

Captain Cook
42

Trekking
across Australia
44

The Amazon
46

Deep inside Africa
48

Livingstone
and Stanley
50

The North Pole
52

Race to the
South Pole
54

Seas, Summits
and Skies
56

Blasting into Space
58

Glossary
60

For More Information
62

For Further Reading
63

Index
63

Introduction

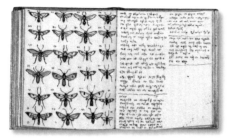

▲ SCIENCE
Explorers in the 18th century set out to record the world they found. Illustrators were taken on expeditions to catalogue the wildlife of the islands that they discovered.

▼ KEY DATES
The panel charts voyages of discovery on Earth, from the first sea voyages in the Mediterranean 3,500 years ago to the conquest of the South Pole in 1911.

Ever since the first people walked on the Earth they have explored the world they lived in. In the beginning, hundreds of thousands of years ago, this was to hunt and gather food; later on it was to find new pastures for their animals. But food was the reason for exploring, and people rarely went far from the place they were born.

When the first civilizations began in the Middle East people began to live in towns and cities. Farmers grew crops and traders bought and sold goods that were not available in their own area. It was these intrepid merchants, from ancient civilizations like Phoenicia and Egypt, who were the first explorers. The Phoenicians sailed out from the Mediterranean into the North Atlantic, while the Egyptians ventured south into the Indian Ocean, looking for opportunities to trade and to establish permanent trading posts or colonies.

Throughout history trade has remained the driving force of discovery. It was the search for a new trade route to China and India that sent da Gama round Africa into the Indian Ocean, and Columbus across the Atlantic. Explorers like Hudson and Bering braved the Arctic Ocean trying to find ways around the top of America and Siberia. And it was trade that sent Magellan round the world and European sailors into the Pacific

▼ CONVERSION
Many European explorers set out to convert the local people they encountered to Christianity.

EUROPE

Phoenician galley

c.1400 BCE Phoenician sailors explore Mediterranean.

c.330 BCE Pytheas sails from France to Thule.

CE 300s First Barbarian invasions of Roman Empire.

1419 Henry "The Navigator" establishes school of navigation in Portugal.

Pilgrims

ASIA

Caravanserai on the Silk Road

c.500 BCE Silk Road opened.

138 BCE Chang Ch'ien, from China, travels into Central Asia.

CE 399 Fa Hsien travels from China to India and Sri Lanka.

1099 First Christian crusaders visit Palestine.

1271–95 Marco Polo visits China.

1325–54 Ibn Battuta, from Algeria, travels round Islamic world.

1405–33 Zheng He, from China leads expeditions to Southeast Asia.

1498 Da Gama, from Lisbon, Portugal, sails to India.

1549 Xavier, a Spanish Jesuit, goes to Japan as a missionary.

1594–97 Barents, a Dutch mariner, explores Arctic Ocean.

1725–29 Bering, from Denmark, crosses Siberia.

1734–42 Teams of explorers map Siberian coast and rivers.

1878–79 Nordenskjöld, from Finland, discovers the Northeast Passage.

Inuit igloo

AFRICA

Timbuktu

c.1490 BCE Egyptians sail to Punt.

c.600 BCE Phoenician fleet sails round Africa.

c.500 BCE Hanno, from Carthage in modern Tunisia, explores coast.

CE 1480s Portuguese cross the Equator and sail round Cape of Good Hope.

1768–73 Bruce, from Scotland, searches for source of the River Nile and discovers Lake Tana.

Introduction

▶ NAVIGATION
The first explorers had little to help them navigate apart from the positions of the Sun, Moon and stars and had to stay close to land. The development of instruments like the magnetic compass, astrolabe and quadrant made navigation easier and more exact.

Ocean to the rich spice islands of Southeast Asia.

Explorers also set out seeking fame and fortune and for political advantage – to conquer new lands for their king and country. Many European explorers travelled out of religious conviction, attempting to convert other races to Christianity. But by the 18th century, it was scientific curiosity that sent Cook into the Pacific Ocean and Bates into the Amazon rainforest.

Today there are few places left on Earth that have not been fully explored. There are unexplored mountains in Tibet and the ocean floor remains largely undiscovered, but now our attention has turned to the skies and space. Unmanned space probes explore the planets of our solar system and the wider reaches of our galaxy of stars, looking for life on other planets. Exploration has come a long way since those early sailing ships left the shores of Phoenicia and Egypt more than 3,500 years ago.

▲ TRADE
The Dutch East India Company was a powerful trading organization. By 1700 they had control of the valuable cinnamon, clove and nutmeg trade in the East. They established trading posts throughout Asia and ruled what is now called Indonesia.

1795–1806 Park, from Scotland, explores the River Niger.

1841–73 Livingstone, from Scotland, explores southern and central Africa.

1844–45 Barth, from Germany, explores the Sahara Desert region.

1858-63 Englishman John Speke discovers the source of the River Nile.

1874–77 Stanley, from Wales, sails down Congo River.

Stanley's hat

AMERICA and the ARCTIC

CE 980s–90s Vikings settle in Greenland and explore parts of North America.

1492 Columbus, from Italy, finds the West Indies.

1497 Cabot finds Newfoundland.

1502 Amerigo Vespucci, from Spain, explores South America.

1513 Balboa sights Pacific Ocean.

1519–33 Spanish conquer Aztecs.

1535–6 Cartier, from France, journeys up St Lawrence River.

1603–15 Champlain explores Canada and founds Quebec.

1610–11 Englishman Henry Hudson searches for Northwest Passage.

1680–82 La Salle, from France sails down Mississippi River.

1800s Several scientific expeditions explore the Amazon.

1804–06 Americans Lewis and Clark explore Louisiana Territory.

1903–06 Amundsen, from Norway, finds the Northwest Passage.

1908 Peary, from America, reaches North Pole.

Lewis and Clark

AUSTRALASIA and the ANTARCTIC

c.1000 BCE Polynesians settle in Tonga and Samoa.

Boomerang

CE 400 Polynesians reach Easter Island and Hawaii.

c.1000 Maoris settle in New Zealand.

1520–21 Magellan crosses Pacific on his round-the-world voyage.

1605 Jansz explores Queensland.

1642–43 Dutchman Tasman discovers New Zealand.

1770 Cook lands in Australia.

1828–62 Interior explored.

1911 Amundsen reaches South Pole.

Egyptians, Phoenicians and Greeks

People have been exploring the world since ancient times. The earliest civilizations grew up in the Middle East thousands of years ago. Merchants began to trade with far-off cities so that they could get hold of goods that were not available in their own land. Gold, spices and craftworks were bought and sold. The easiest way to make long journeys to other countries was by sea. The traders had no maps to guide them, so they had to discover the best routes for themselves. They soon learnt about the winds and sea currents that would help their voyages, and which seasons were best to travel in.

The ancient Egyptians lived along the banks of the River Nile. They had plenty of food and other goods, so traders did not venture very far. But eventually the traders wanted to find new markets, and this tempted them to explore further afield. They started to sail ships out into the Mediterranean and the Red Sea.

In 1490 BCE, Queen Hatshepsut of Egypt ordered a fleet to sail down the Red Sea in search of new lands. The fleet reached a place called Punt (modern Somalia or somewhere further down the coast of East Africa). The sailors returned with ivory, ebony, spices and myrrh trees – a present from the people of Punt. Other expeditions explored the interior of North Africa.

Phoenician sailors began to explore the Mediterranean Sea in about 1400 BCE. The Phoenicians lived in cities along the coast of what is now Lebanon, at the eastern end of the Mediterranean. They were skilled seafarers and soon started to establish prosperous trading colonies throughout the region. One Phoenician fleet even sailed round Africa on behalf of an Egyptian pharaoh. In 500 BCE a man called Hanno sailed from Carthage, a Phoenician colony in North Africa, as far as modern Senegal, a journey of 2,500 miles. Other Phoenician traders sailed to Britain, buying tin in Cornwall.

▲ BABOON
The Egyptians brought live baboons and cheetahs back from Punt, as well as leopard skins.

▶ A PHOENICIAN SHIP
Phoenician ships were short, broad and strong. They were built from cedar, which grew on the mountain slopes of Phoenicia. A single sail and oars powered the ship along.

SAILING THE MEDITERRANEAN
Phoenicia had little arable land, and so in 1400 BCE its people turned to the sea for a living. They became excellent seafarers, sailing great distances in search of new markets. They established colonies as far away as North Africa and Spain. Egyptians and Greeks also began to explore by sea.

◀ PHOENICIAN TRADERS
The Phoenicians traded grain, olive oil, glassware, purple cloth, cedar wood and other goods throughout the Mediterranean area. They were rather like travelling shopkeepers.

▼ MUREX SHELL
One of the most precious items traded by the Phoenicians was purple cloth. The dye for the cloth was made from murex shells. Up to 6000 shells were crushed to make one pound (450 grams) of dye.

▶ PHOENICIAN GLASS
The Phoenicians were good at making glass items, such as vases and jewellery. Sand and soda were mixed to make a paste, which was colored with pigment and fired at a high temperature.

EGYPTIANS, PHOENICIANS AND GREEKS

▼ EGYPTIAN PORT
In Egypt, shallow-bottomed boats made of reeds, with a single sail, carried goods and passengers along the River Nile. After about 2700 BCE the Egyptians began to build wooden boats, which were stronger and could cross seas to foreign lands.

The Greeks also founded colonies throughout the Mediterranean. The Phoenicians were their great rivals, because they were so successful at trading by sea. Greek merchants wanted some of the business for themselves.

In 330 BCE, an explorer called Pytheas sailed to Britain, possibly to try to get access to the profitable tin trade.

▶ PYTHEAS
One of the most amazing voyages of ancient times was made by a Greek astronomer called Pytheas. In 330 BCE he set sail from Marseille in southern France, which was a Greek colony. He headed round Spain and then north to the British Isles, where he reported that the local people were friendly. Pytheas continued his voyage further north to the land of Thule. Thule was probably Norway or Iceland. Pytheas noted that in Thule the sun never set. (In these countries it does not get dark in summer.)

Key Dates

- 1490 BCE Egyptians sail to Punt.
- 1400 BCE Phoenician traders explore the Mediterranean Sea and the eastern Atlantic Ocean.
- 1000 BCE First Phoenician colony established on Cyprus.
- c.800 BCE Greeks set up colonies in the eastern Mediterranean.
- 814 BCE Phoenicians found Carthage in North Africa.
- c.600 BCE Phoenician fleet sails round Africa.
- 500 BCE Hanno explores the coast of West Africa.
- 330 BCE Pytheas sails to Thule.

Europe and Asia

▲ SILKWORM
A silkworm feeds on a mulberry leaf. The Chinese began cultivating silkworms for silk over 4,500 years ago.

In ancient times there was not much contact between Europe and Asia. In Europe the flourishing trading empires of the Phoenicians and Greeks were centred on the Mediterranean Sea. In eastern Asia the Chinese had their own trading centers. In between the two continents were the deserts, mountains and arid plateaux of central Asia.

The Chinese were famous for making beautiful silk fabric. A few hardy traders made the long journey between Europe and China along a route known as the Silk Road. They bought bales of silk from Chinese merchants to take back with them. There are records of Chinese silk being sold in the Greek city of Athens as early as 550 BCE.

Two hundred years later, King Alexander of Macedon (later known as Alexander the Great) invaded the huge Persian Empire, which extended into central Asia. Many scholars and historians went with him. They began to explore the vast regions that Alexander had conquered, and learnt a lot about them.

When Alexander died his empire broke up. But links between Europe and Asia became stronger over the next century. The empires of different rulers lay all along the Silk Road. The Romans controlled Europe, the Parthians ruled Persia and the Kushans dominated central Asia. In China the country became united for the first time under the first Ch'in emperor in 221 BCE. These four empires spanned the length of the Silk Road, and for more than 400 years there was uninterrupted trade between East and West. Few Roman merchants visited China, but a wide variety of goods flowed along the Silk Road in both directions.

There was also a thriving sea trade across the Indian Ocean between Egypt and India, and from there on to China. This too helped to increase international trade and contacts.

The Silk Road was also important in linking the different Asian empires

▶ TRADERS
The Silk Road was a busy route. Merchants from Europe, the Middle East, central Asia and China used it on trips to buy and sell goods. However, not many of them ever travelled along its entire length.

FROM WEST TO EAST
The major trade route between China and Europe was known as the Silk Road, because Chinese silk was brought along it by traders returning to Europe. In exchange, China received gold and silver, cotton and a wide variety of fruits and other produce.

▼ ALEXANDER THE GREAT
Alexander was only 20 when he succeeded his father as king of Macedon in 336 BCE. By the time of his death 13 years later, he had conquered an empire that stretched from the Adriatic Sea in southern Europe to the mouth of the River Indus in India.

◀ BEASTS OF BURDEN
Donkeys, horses and two-humped Bactrian camels were all used on the Silk Road. They worked hard carrying the traders and their cargoes.

▲ JADE
Jade was the most precious substance known to the Chinese. They carved it into elaborate and intricate ornaments and utensils, such as this brush-washer.

Europe and Asia

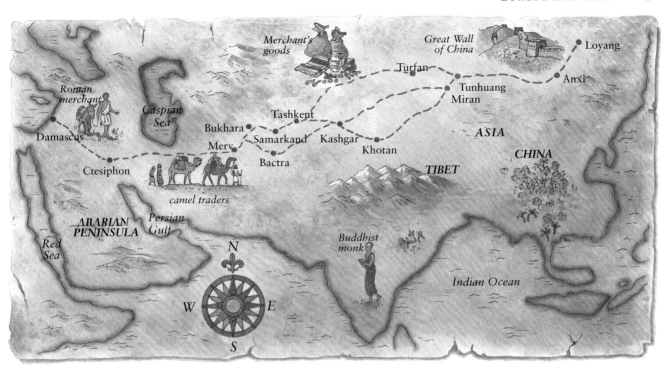

together. Buddhist monks brought their religion from India into China in about CE 100. Chinese explorers journeyed into neighboring countries, which helped to strengthen religious and trading ties between them. Chang Ch'ien, a court official, travelled into central Asia in 138 BCE. A monk called Fa Hsien visited India and Sri Lanka in CE 399. However, by CE 400 these ties had weakened. Civil war broke out in China and barbarian invaders and nomads from central Asia overran the Silk Road. By CE 450 the links between East and West were broken.

▲ **THE SILK ROAD**
The Silk Road started in the Chinese capital of Loyang and ran westwards across northern China and central Asia to Ctesiphon on the River Tigris in southwest Asia. It later continued to the Mediterranean. It was not a single road, but a series of well-marked routes. Traders using them were fairly safe from attacks by robbers.

▶ **CARAVANSERAI**
Traders on the Silk Road stopped each night in one of the many caravanserais, or inns, that lay along the route. Here they relaxed, had a meal and exchanged information and gossip with their fellow traders. Their animals were fed and had a good rest before the next day's journey.

Key Dates

- c.500 BCE Silk Road established for trade from China to Europe.
- 334 BCE Alexander of Macedon conquers Persian Empire.
- 221 BCE China united.
- 138 BCE Chang Ch'ien travels from China into central Asia.
- CE 100 Buddhism reaches China.
- CE 166 Roman traders in China.
- CE 220 After lengthy civil war China splits into three parts.
- CE 399 Fa Hsien travels from China to India and Sri Lanka to study Buddhism.

The Invasion of Europe

THE CITY OF ROME started off as a small cluster of villages in the hills around the River Tiber in central Italy. By 500 BCE, it had grown into the most powerful city in Italy. Rome declared war on its main rival, the Phoenician city of Carthage. Carthage was defeated, and the Romans built an empire that spanned the Mediterranean. By the reign of Emperor Trajan (CE 98–117), the Roman Empire stretched from Portugal to Mesopotamia, and from the Sahara Desert to the borders of Scotland.

The long eastern frontier of the Romans' vast empire was easy to invade. In CE 300, the Huns from the vast steppes of central Asia began to head westwards into southeast Europe, in search of new pastures. This pushed the Germanic tribes living there (including the Goths, Visigoths, Vandals, Alans and Franks) across the Roman frontier.

At first the various tribes lived peacefully in the Roman Empire. But in CE 378 fighting broke out, and the Visigoths defeated the powerful Roman army in the Balkans. From there the Visigoths moved further into the Roman Empire. They successfully attacked Rome in CE 410 and continued westwards to settle in Spain. The Franks and Vandals crossed the River Rhine and invaded Gaul (France). The Vandals kept going south until they reached North Africa. Finally the Huns, led by Attila, arrived in Europe.

▲ GOTHIC ART
This illuminated manuscript from the 1100s is by the Goths. They were the first Germanic people to become Christians. Many years later an ornate style of art developed based on original Gothic designs.

▼ ROME DESTROYED
In CE 410 the Visigoths captured and plundered Rome, killing many of its inhabitants. The Vandals did the same in CE 455. But Rome continued as a thriving city until CE 546, when the Ostrogoths captured it and expelled its entire population, leaving it in ruins.

THE DARK AGES

The period after the fall of the Roman Empire is often called the Dark Ages. This is because Europe became a poorer place. People managed to grow enough food, but trade and commerce declined rapidly. Europe seemed to be taking a step backwards from the great culture and prosperity enjoyed by the Romans. However, learning and scholarship continued in monasteries everywhere.

Christianity became the main religion of Europe and the new rulers of Europe soon developed cultures of their own.

▲ ENAMELLED BROOCH
Many of the barbarian invaders were skilled craftworkers, as can be seen from this brooch.

▶ LINDISFARNE
In monasteries such as Lindisfarne Priory, in northeast England, monks kept education alive during the times known as the Dark Ages.

▲ MAUSOLEUM
The barbarian invaders of Europe soon adopted Roman customs. This mausoleum was built in the Roman style in CE 526 in Ravenna, northern Italy, as the burial place of Theodoric, an Ostrogoth chief.

The Romans managed to defeat Attila in CE 453 with the help of friendly Germanic tribes, but by then their empire was weak and exhausted. In CE 476 the last Roman emperor was removed from power. Rival invading tribes fought over the remains of the Roman Empire. The Romans called the invaders barbarians, but many were educated people looking for a safer land to live in.

The old Roman Empire was swept away, but the tribes of Europe remained disunited for several hundred years. Then in CE 800 Charlemagne, a Frankish king from northern Europe, was crowned "Holy Roman Emperor" by the pope in Rome. By that time Christianity was the main religion across Europe. Between them, the pope and Charlemagne brought some unity to the peoples of western Europe.

▲ ROMAN RUINS
After the collapse of the Roman Empire many of its impressive buildings, such as this forum, fell into disuse. The stone was removed to construct new buildings and the roofs were left to rot and fall in.

▲ CHARLEMAGNE
In CE 768 the Frankish leader Charlemagne began to conquer a vast empire that included modern France, Germany, the Low Countries and Italy. The new empire reunited western Europe for the first time since the fall of the Roman Empire.

Key Dates

- 264–146 BCE Rome overthrows Carthage after three wars.
- CE 300s Germanic tribes begin to enter Roman Empire.
- CE 378 Visigoths defeat Roman army in the Balkans.
- CE 410 Visigoths ransack Rome.
- CE 455 Vandals plunder Rome.
- CE 476 Roman emperor deposed.
- CE 546 Ostrogoths destroy Rome.
- CE 768 Charlemagne, Frankish leader, begins to reunite Europe.
- CE 800 Charlemagne is crowned emperor by the pope in Rome.

The Vikings

▲ BROOCH
Lavishly decorated clasps and brooches made of bronze, silver and gold were used to hold clothes in place.

▼ VIKING SAILORS
Much of the Vikings' homeland was mountainous, with few roads. They used their boats to travel the fjords and the open seas.

The Vikings seemed to come from nowhere. Setting out from Norway and Denmark, they suddenly became a frightening force that dominated the northern seas, from the Atlantic Ocean to the Black Sea. They struck terror into all those they met on land. The records kept by Christian monks describe them as ruthless fighters who plundered towns and slaughtered all the inhabitants. Few towns managed to hold off these ferocious invaders.

The name "Viking" means "men of the creek" and they came from the fjords and lowlands of Scandinavia, in the north of Europe. Although the Vikings have a reputation for cruelty, they were a talented people. They were skilled shipbuilders and navigators, and excellent engineers and craftworkers. They had a rich tradition of myths and legends, and had worked out a fair system of rules for the way their people lived.

In about CE 790, parties of Vikings began to leave their homeland, launching their boats into the open sea. No one is quite sure why they began to do this. Some historians believe that overcrowding in the country drove out the younger sons who had nothing to inherit from their fathers. Or perhaps the climate was getting colder and harvests often failed, leading people to search for new sources of food.

Vikings from Norway and Denmark crossed the North Sea to raid Britain, Ireland and northern France. They ventured across the North Atlantic to Iceland, Greenland and the east coast of America. Vikings from Sweden, however, preferred trading with other countries to conquering them. They sailed east, across the Baltic Sea and down the rivers of Russia to the Black

THE VIKING WAY OF LIFE
There were three classes of Vikings – slaves, who did most of the work, freemen and nobles, who were the rulers. Nobles had to obey rules made by the Thing, a local assembly where freemen could discuss these rules. But by about 1050, powerful kings ruled most Viking lands. The Things were no longer so important and their role gradually declined.

◀ CLOTHES
Clothes were made from wool or flax spun on a vertical loom. Women wore a long dress with a shorter tunic on top. Men typically wore trousers, a shirt, a tunic and a cloak.

▶ LEIF ERIKSSON
Vikings living in the colony of Greenland heard stories of a flat land to their west covered with trees. In CE 992 Leif Eriksson set out due west, and found what was probably Baffin Island. He then sailed south past Labrador and Newfoundland, in eastern Canada, to a place he named Vinland, the "land of wine," as he found so many shrubs and wild berries there.

◀ RUNES
The Vikings used an alphabet system based on runes, which were usually carved in wood or on pieces of stone. Calendars, bills, accounts and even love messages were all written in runic script.

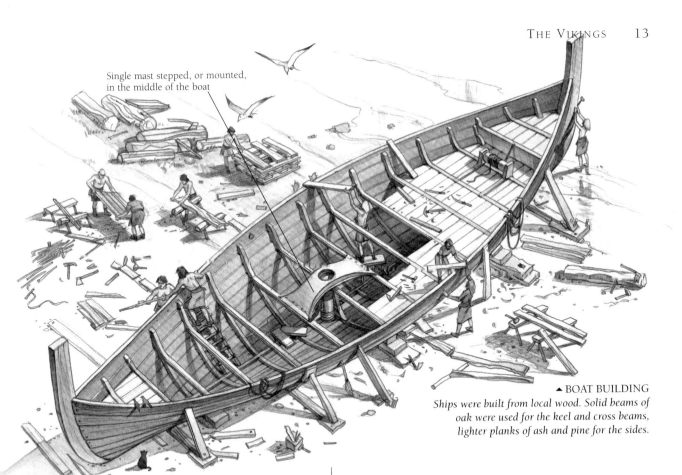

Single mast stepped, or mounted, in the middle of the boat

▲ BOAT BUILDING
Ships were built from local wood. Solid beams of oak were used for the keel and cross beams, lighter planks of ash and pine for the sides.

Sea and the Mediterranean. They even reached the city of Baghdad on the River Tigris in modern Iraq.

At first the Vikings plundered the lands they visited, taking their booty home. But gradually they started to establish trading posts in places such as Dublin in Ireland and Kiev in Ukraine. Soon they began to marry the local people and settle down. Many converted to Christianity. The Viking raids were over.

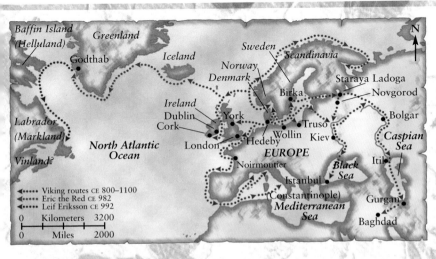

▲ VIKING EXPEDITIONS
Viking traders from Sweden sailed down the rivers of Russia. Others sailed round the Atlantic coast into the Mediterranean. Eric the Red was expelled from the Viking colony on Iceland and sailed to Greenland and Leif Eriksson voyaged to Vinland.

Key Dates

- CE 793 Vikings attack Lindisfarne Priory in northeast England. It is their first major raid on England.
- CE 815 Vikings from Norway settle in Iceland.
- CE 841 Foundation of Dublin.
- CE 855 Vikings sail up the River Seine in France and raid Paris.
- CE 911 Vikings settle in Normandy, France.
- CE 982 Eric the Red begins his epic voyage to Greenland.
- CE 992 Leif Eriksson leaves Greenland and sails to Vinland.

The Polynesians

THE ISLANDS OF THE South Pacific were uninhabited until about 3,000 years ago. Then the first Polynesians arrived to live there. We do not know much about these people. Historians think that they originally came from Asia or America.

Over the next 2,000 years the Polynesians slowly spread out across the vast South Pacific Ocean. They sailed north to Hawaii, east to Easter Island and, finally, south to New Zealand. They were probably the greatest explorers and navigators in history. When Europeans first visited the region in the 1500s, they got a surprise. They could not believe that the Polynesians, who they thought were a very primitive people, could have developed such advanced skills.

The immense Pacific Ocean is scattered with islands, but these make up only a minute part of its total area and lie hundreds of miles apart from each other. The rest is open sea, and it is easy to sail for days without sighting any land. The Polynesians did not have any maps or modern navigation equipment, but they successfully explored the entire ocean in their sturdy canoes. They settled on almost every island, finding them by following migrating birds and by watching changes in wind direction and wave pattern.

The Polynesians gradually built up a detailed knowledge of where each island was and how they could find it again in the future, using the Sun, Moon and stars as navigation aids. They gave each island its own "on top" star. Sailors knew that when this was directly over their boat, they were on the same latitude as the island. Using the position of the Sun, they sailed due east or west until they reached land. Sirius, for example, was the "on top" star for Tahiti.

All this information was passed down through the generations and recorded on a chart made of palm sticks tied together with coconut

▲ PELE GOD
Polynesians made statues, as here, of dead ancestors, as they thought their spirits became gods.

▶ GIANT STATUES
Easter Islanders erected 600 giant carved statues across their small island between CE 1000 and 1600. No one knows what these statues were for, or how the islanders managed to move and erect the huge stones.

ASIAN OR AMERICAN?
Some historians think that the Polynesians originally came from Southeast Asia, but there are many similarities between the cultures of Polynesia and Peru. One modern explorer from Norway called Thor Heyerdahl set out to prove that Polynesians could have come from South America. He built a raft like those used by early settlers and sailed from Peru to the South Pacific.

▼ THOR HEYERDAHL
Thor Heyerdahl was born in 1914 and studied zoology and geography at university. He became fascinated by Polynesia and lived for two years in Tahiti.

◀ KON-TIKI
Thor Heyerdahl's raft was called *Kon-Tiki* after the Peruvian sun-god. The god was believed to have migrated to the Pacific islands. The raft measured 15 yards (13.7 m) long and 6 yards (5.5 m) wide, and was made of balsawood and bamboo.

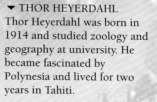

▲ SWEET POTATOES
South Pacific islanders ate sweet potatoes. The plant comes from the Americas and Heyerdahl thought it might have been taken to Polynesia by settlers from America.

THE POLYNESIANS 15

▼ POLYNESIAN BOAT
Polynesian canoes were up to 33 yards (30m long). They were built with two hulls or a single hull and an outrigger. The sails were made from coconut-palm leaf matting stitched together.

Canoe steered by single oar

Outrigger

Main hull

▲ PACIFIC ISLANDS
There are about 20,000 islands in the Pacific Ocean. Most are either high volcanic peaks or low coral reefs. Apart from New Zealand, the vast majority are small, some only a few miles across.

fiber. The framework of sticks represented distance, and shells threaded on the sticks showed where the islands were. The Polynesians used these simple but effective charts to make accurate voyages across vast expanses of ocean. They took colonists and supplies to newly discovered islands and brought back fish and other goods.

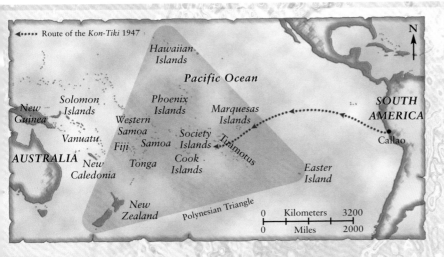

▲ THOR HEYERDAHL'S VOYAGE
In 1947, *Kon-Tiki* set sail from Peru. It steered westwards, making use of the winds and sea currents. After a voyage of 4,287 miles (6,900 km), lasting 101 days, Thor Heyerdahl reached the Tuamotu archipelago in the South Pacific.

Key Dates

- 1000 BCE Polynesians begin to settle in Tonga and Samoa.
- 150 BCE Settlers leave Samoa for Marquesas Islands.
- CE 400 Polynesians reach Easter Island in the east and the Hawaiian Islands in the north.
- CE 1000 Polynesian Maoris settle in New Zealand.
- CE 1000–1600 Statues built on Easter Island.
- CE 1947 Thor Heyerdahl's *Kon-Tiki* expedition from Peru to the South Pacific.

Crossing the Deserts

▲ THE DESERT
The deserts of North Africa and Arabia are a mixture of rocky plains, sand dunes and high mountains.

In CE 622 Arabia gave birth to a new religion called Islam, which soon became the main religion there. Islam was founded by the Prophet Mohammed, who died in CE 632. Within a few years an Islamic Empire stretched from the Atlantic Ocean across North Africa and the Middle East to the borders of India. Much of the Islamic world was desert, including the Sahara Desert in North Africa, which is the biggest in the world.

The Arabs dominated the Islamic Empire, conquering other lands. Soldiers and missionaries spread Islam to all the people they conquered. The Arabs were great travelers. Merchants and traders followed the soldiers into the new lands in search of markets to exploit and goods to trade. Scholars and scientists travelled to increase their knowledge of the world. Every Muslim (a believer in Islam) also made a once-in-a-lifetime pilgrimage to the holy city of Mecca in Arabia. As a result, towns and cities were bustling with travelers stopping to rest for the night after a long day's journey. Markets and public places were full of people telling stories about faraway cities and strange lands they had visited.

▲ MUSLIM TRAVEL
Muslim travelers felt at home wherever they went in the Islamic world, since everyone spoke the common language of Arabic and all were followers of Islam. Travelers got a warm welcome at every town.

MUSLIM TRAVELERS
There were many reasons why the Muslims were great travelers. One was because of Islam. It was the duty of all Muslims to make a pilgrimage to the holy city of Mecca. People also travelled on business, or because they were curious about other places.

▲ MUSLIM PRAYER CASE
Muslims carried verses from their holy book, the Koran, in prayer cases. These were often carved and decorated.

◀ SALT
Salt occurs naturally throughout the Sahara Desert, and was much prized as a food preservative. Camel trains were used to bring the mineral to the Mediterranean for shipment to Europe.

▼ IBN BATTUTA
Ibn Battuta was born into a wealthy, educated family in Tangier, North Africa, in 1304. After a pilgrimage to Mecca in 1325, he was inspired to devote his life to travel. After his final journey in 1352–53, he recorded his adventures in a book called the *Rihlah,* which means "Travels." He died in 1368.

Crossing the Deserts 17

◀ ARAB DHOW
The main type of ship used by the Arabs for their travels was the dhow. It had one or two masts, and a triangular sail bent on to a spar. This was called a lateen sail.

▲ SHIP OF THE DESERT
Camels can travel for many days without food or water. Fat stored in their humps gives them energy when food is scarce.

Many Muslims left records of their travels, such as Ibn Battuta, who was the greatest traveller of his age. The geographer Al Idrisi (1100–65) used his explorations to produce a book on geography and two large-scale maps of the world. Al Idrisi even sent out scientific expeditions to explore northern Europe and other areas unfamiliar to the Arabs.

At first Arabs avoided the Indian Ocean, calling it the "Sea of Darkness." But then they developed the dhow, a versatile ship that could carry large cargoes, but which only needed a small crew. Astrolabes, quadrants and other devices were used to navigate. At that time, most European people thought that the world was flat and that it was unsafe to venture far beyond land. The Arabs were sailing across the Indian Ocean to purchase silks and spices from India, Indonesia and China, and travelling vast distances as traders, pilgrims or adventurers.

▼ THE TRAVELS OF IBN BATTUTA
Ibn Battuta spent a total of 28 years travelling to find out more about the world. He explored the entire Islamic Empire, as well as much of Europe, South-east Asia and China. He covered a total distance of 75,000 miles (120,700 km).

Key Dates

- CE 622 Mohammed and his followers flee persecution in Mecca and go to Medina, starting the Islamic religion.
- CE 632 Death of Mohammed.
- CE 634–50 Muslims conquer Middle East.
- CE 650 The Koran (the sacred book of Islam) is written.
- CE 712 Islamic Empire extends east to Spain and west to India.
- CE 1200s The dhow sailing boat is developed.
- CE 1325–53 Ibn Battuta travels round the Islamic world.

The Chinese Empire

▲ PORCELAIN
The Chinese invented a new type of pottery called porcelain. It is very fine and hard, and light can shine through it.

During the 1200s, a new threat emerged from the empty, hostile grasslands or steppes of central Asia. It was the Mongols, a nomadic people who were skilled horsemen and warriors. They conquered most of Asia and eastern Europe in a series of brilliant military campaigns led by their ruler, Genghis Khan, and his successors. Their empire stretched from the eastern frontier of Germany to Korea, and from the Arctic Circle to the Persian Gulf. It was the biggest empire the world had ever seen.

Although the Mongols had a reputation for extreme violence, they kept strict law and order throughout their empire and encouraged trade. The main trade link between Europe and Asia – the Silk Road – had fallen into disuse after the collapse of the Roman Empire and the Han dynasty in China. The Silk Road was now in the Mongol Empire. The Mongols made sure that the road was safe to use.

Two Venetian merchants, Niccolò and Maffeo Polo, were among the first traders to travel its entire length. The Mongol ruler of China, Kublai Khan, welcomed them to his court, as he was fascinated by the foreigners and the mysterious lands they came from. When the Polos returned home, they promised him that they would serve as his envoys to the pope in Rome and arrange for 100 theologians to go to China to discuss Christianity with Mongol philosophers. Kublai Khan gave them a golden tablet inscribed with the imperial seal to guarantee them good treatment and hospitality while in Mongol territory.

When the Polo brothers returned to China, they took Niccolò's son Marco with them. He served Kublai Khan for the next 20 years and travelled throughout

▲ VENICE
The port of Venice in Italy was the richest city in Europe in the 1200s. It controlled much of the trade in the Mediterranean. It was from here that the Polo brothers set out on their travels.

CHINESE VOYAGES

In 1368, nearly a century after Marco Polo's amazing reports of the court of Kublai Khan, the Mongols were thrown out of China. To try and restore China's prestige in Asia, the new Ming dynasty sent Admiral Zheng He (1371–1433) on a series of seven diplomatic journeys through the whole region.

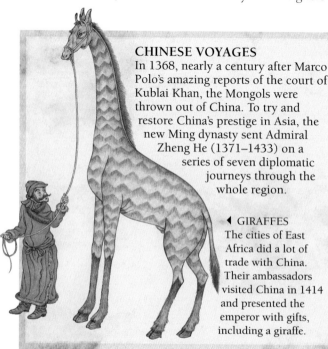

◀ GIRAFFES
The cities of East Africa did a lot of trade with China. Their ambassadors visited China in 1414 and presented the emperor with gifts, including a giraffe.

▲ CHINESE JUNKS
Zheng He commanded a fleet of ocean-going junks. These were two-masted ships that could carry large cargoes. Some junks were five times larger than European ships of the time.

The Chinese Empire 19

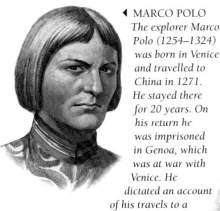

◀ MARCO POLO
The explorer Marco Polo (1254–1324) was born in Venice and travelled to China in 1271. He stayed there for 20 years. On his return he was imprisoned in Genoa, which was at war with Venice. He dictated an account of his travels to a fellow prisoner. Il milione (The Travels) was read throughout Europe.

eastern Asia. When Marco Polo returned to Europe, he wrote a book about his epic journeys. He described the fabulous court of Kublai Khan, and he praised the artistic and technological achievements of Mongol China. However, some modern scholars believe that Marco Polo did not go to China, but wrote his book after reading other people's reports.

▲ KUBLAI KHAN
The Mongol ruler of China, Kublai Khan, was a highly intelligent and civilized man. His summer palace at Shangdu, where he welcomed the Polos to China, and his court in Cambaluc (Beijing) were both magnificent buildings.

▼ MARCO POLO'S ROUTE
In 1271 the Polo family set out from Europe along the Silk Road, taking over three years to reach China. There they stayed for 20 years, while Marco travelled round the vast Mongol Empire. They returned to Europe across the Indian Ocean in 1295.

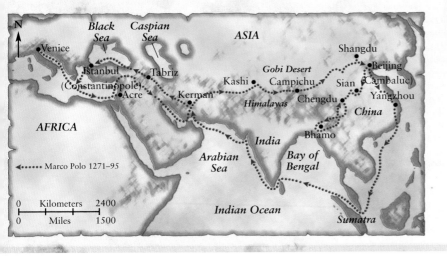

Key Dates

- 1211 Genghis Khan begins invasion of China.
- 1234 Mongols overrun northern China.
- 1260–94 Mongol leader Kublai Khan rules China.
- 1261–69 Polo brothers travel to China.
- 1271–95 Marco Polo travels throughout the Mongol Empire.
- 1368 Mongols thrown out of China by the Ming dynasty.
- 1405–33 Zheng He leads seven expeditions to Southeast Asia and Indian Ocean.

Travelers in Europe

▲ CITY OF DANZIG
The ship on Danzig's official seal shows membership of the Hanseatic League, a network of trading cities in northern Europe.

TODAY WE KNOW OF Europe as a busy place with a huge population. However, a thousand years ago Europe was a very different place indeed.

In the year 1000 the total population of Western Europe was fewer than 30 million, which is about half that of modern France. Only a handful of cities, including Paris, Milan and Florence, had more than 40,000 people. Most had fewer than 10,000 inhabitants. Roads were rough and uneven, and much of the countryside was covered in thick forests. Bandits lay in wait to rob travelers. Yet despite these problems, a surprising number of people ran the risk of getting lost or being robbed and took to the roads.

Pilgrims journeyed in great numbers to holy sites, such as the shrine of St James in Compostela, northern Spain. Some went even further afield, to Jerusalem in the Middle East. Armies of Crusaders assembled to reconquer the Holy Land from its Muslim occupiers. National armies marched off to fight wars on behalf of their kings. Merchants and traders travelled from town to town, buying and selling goods at the increasing number of trade fairs held in northern France, Germany and Flanders (Belgium). Government and church administrators moved from town to town on official business. Scholars often passed from one university to another for their studies. Craftworkers and builders made their way to the cities where new cathedrals were being built.

▼ PILGRIMAGES
Pilgrims travelled to the great religious shrines in large groups, telling each other stories and singing songs to pass the time.

THE CRUSADES
Between 1095 and 1444, armies of Christian knights went to Palestine in the Muslim Empire. They intended to secure the Christian holy places against Muslim control. The First Crusade successfully captured Jerusalem. In 1291, the last Crusader stronghold was lost. Later Crusades all ended in failure.

◀ THE CRUSADERS
Crusading knights and soldiers were inspired by religious devotion. They also followed a code of conduct called chivalry, which meant that they pledged to be brave, loyal to their lord, and protective of women.

▲ RICHARD I
Many kings and princes joined the Crusades. Richard I of England (ruled 1189–99) took part in the Third Crusade.

▼ HEIDELBERG
Heidelberg, and other fortified towns throughout Germany, were good centers for recruiting men for the Crusades. All types of people went, commoners and knights, to spread Christianity through the Holy Land.

Travelers in Europe

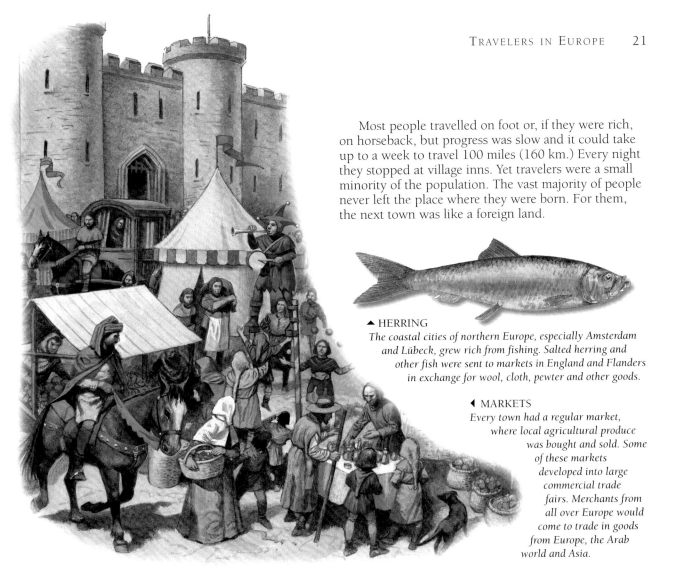

Most people travelled on foot or, if they were rich, on horseback, but progress was slow and it could take up to a week to travel 100 miles (160 km.) Every night they stopped at village inns. Yet travelers were a small minority of the population. The vast majority of people never left the place where they were born. For them, the next town was like a foreign land.

▲ **HERRING**
The coastal cities of northern Europe, especially Amsterdam and Lübeck, grew rich from fishing. Salted herring and other fish were sent to markets in England and Flanders in exchange for wool, cloth, pewter and other goods.

◄ **MARKETS**
Every town had a regular market, where local agricultural produce was bought and sold. Some of these markets developed into large commercial trade fairs. Merchants from all over Europe would come to trade in goods from Europe, the Arab world and Asia.

▼ **KRAK DES CHEVALIERS**
The Crusaders built castles throughout Palestine to secure their conquests against Muslim invaders. The most impressive was Krak des Chevaliers, which is in present-day Syria. It eventually surrendered to Muslim armies in 1271.

Key Dates

- 1095 Pope Urban II calls for a Crusade to defend the Church.
- 1099 Crusaders capture Jerusalem and set up Crusader kingdoms throughout Palestine.
- 1187 Muslim leader Saladin retakes Jerusalem and overruns most of the Crusader kingdoms.
- 1189–92 Third Crusade recaptures Acre from Saladin.
- 1241 Hamburg and Lübeck set up the Hanseatic League.
- 1291 Acre, the last Crusader stronghold in Palestine, is lost.
- 1444 Final Crusade.

The Portuguese

▲ DA GAMA
In May 1498 the Portuguese navigator Vasco da Gama (1460–1524) became the first European to reach India by sea.

Portugal is on the extreme west of Europe, facing the Atlantic Ocean. The Portuguese relied on the sea to give them a living. Traditionally, they had fished and traded northwards along the Atlantic coast with France and Britain. But during the 1400s they turned their attention south and started looking at Africa.

The Portuguese wanted to explore Africa for two main reasons. They aimed to convert the Moors (the Muslim people of North Africa) to Christianity. They were also going to search for gold and other riches. To do this they needed better ships than the inshore, open boats they usually sailed. They developed the caravel, which was able to withstand the storms and strong currents out at sea.

Caravels allowed the Portuguese to venture further and further from their own shores. Expeditions boldly set off down the African coast, erecting *padrãoes*, or stone pillars with a Christian cross on the top, to mark their progress. By 1441 they had reached Cape Blanc in what is now Mauritania. By 1475 they had sailed round West Africa and along the coast to the Gold Coast (Ghana) and Cameroon.

▲ CARAVEL
The development of the small but sturdy caravel enabled the Portuguese to leave coastal waters and venture out into the open seas. A caravel was about 22 yards (20 m) long and held a crew of 25.

By now the Portuguese had an extra reason to voyage south. In 1453 the Ottoman Turks had captured the Christian city of Constantinople, which was the gateway to Asia, and closed the Silk Road to China. Europeans needed to find a new way to get to the wealth of the East. In 1482 Diego Cão was the first European to cross the Equator. On his second voyage in 1485–86 he sailed as far south as the Namib Desert. He

NAVIGATION
The first sailors navigated by sailing along the coast from one landmark to the next. Once out of sight of land, they could not do this! Portuguese sailors learnt how to use the positions of the Sun and stars to calculate where they were. With the aid of compasses, astrolabes, quadrants, sand glasses and nocturnals, they were able to navigate over long distances with increasing accuracy.

◄ PRINCE HENRY "THE NAVIGATOR"
Prince Henry (1394–1460) was the son of King John I of Portugal. He was keenly interested in the sea and supported many voyages of exploration. He set up a school of navigation, astronomy and cartography (map-making) to educate captains and pilots. These skills enabled Portuguese sailors to explore the coast of Africa.

► SAND GLASS
Sailors told the time with a sand glass. The sand took 30 minutes to run to the bottom and it was then turned over. To calculate the ship's speed, they floated a knotted rope beside the ship, and worked out how long it took to pass each knot.

▲ NOCTURNAL
The old way of telling the time, by the position of the Sun, did not work at night. The development of the nocturnal during the 1550s solved this problem. By lining it up with the Pole Star and two stars close to it, it was possible to tell the time to within ten minutes.

thought that the African coast was endless, and that there was no way round it towards Asia. But in 1487–88 Bartolomeu Dias proved him wrong when he sailed round the stormy Cape of Good Hope into the Indian Ocean. He was the first Portuguese explorer to enter these waters. Although Dias wanted to go on, his exhausted crew made him turn back. Ten years later Vasco da Gama achieved the Portuguese dream. He rounded the tip of Africa with a fleet of four ships. After sailing up the east coast, he headed across the Indian Ocean. In May 1498 he arrived in the busy trading port of Calicut in the south of India. He had discovered a new route to Asia.

▲ **THE ROUTE TO INDIA**
By slowly mapping the coast of Africa, the Portuguese discovered a route which took them round the Cape of Good Hope to East Africa and then, using the westerly winds, across the ocean to India. Once the coast was mapped, later voyages could take a more direct route.

▶ **USING A COMPASS**
The magnetic compass was developed by both the Chinese and the Arabs. It was first used in Europe during the 1200s. By lining up the compass with the magnetic North Pole, sailors could tell which direction they were sailing in. However, early compasses were often unreliable and were easily affected by other iron objects on board ship. As a result, many ships headed off in the wrong direction. By the time of Henry the Navigator in the 1400s, compasses were much improved.

Key Dates

- 1419 Prince Henry establishes a school of navigation.
- 1420s First voyages south to southern Morocco.
- 1475 Portuguese sailors map the African coast from Morocco to Cameroon.
- 1482 Diego Cão crosses over the Equator.
- 1485–86 Diego Cão sails south to Namibia.
- 1487–88 Bartolomeu Dias sails round Cape of Good Hope.
- 1497–98 Vasco da Gama sails round Africa to India.

Christopher Columbus

▲ COLUMBUS
Christopher Columbus (1451–1506) was born in the Italian port of Genoa. He was named after St Christopher, the patron saint of travelers. His discoveries included Cuba and the Bahamas.

For centuries Europeans believed that the world consisted of just three continents – Europe, Africa and Asia. They thought that the whole of the rest of the world was covered by sea.

The traditional route to Asia had always been overland along the Silk Road. During the 1400s, the Portuguese discovered a way of getting there by sea, sailing south and east round the coast of Africa. Then an Italian called Christopher Columbus worked out that it should be possible to get to Asia by sailing west, across the great Atlantic Ocean.

Columbus devoted his life to finding this sea route to the riches of Asia. At first, people thought that it was a stupid idea and Columbus could not get any support. But in 1492 Queen Isabella of Spain agreed to give him money to make the voyage on behalf of Spain. He set out with three ships in August 1492, and after 36 days landed in what we now call the Bahamas. Sailing southeast, he passed Cuba and Hispaniola (present-day Haiti and the Dominican Republic) before returning home in triumph in March 1493.

▼ LANDING IN AMERICA
When Columbus and his crew landed on Watling Island in the Bahamas, he claimed the island for Spain and renamed it San Salvador "in honour of God who guided us and saved us from many perils."

Columbus was convinced that he had found a new route to Asia. Although he was disappointed that the new lands were not full of gold, he set off again later in the year to confirm the discoveries of his first voyage.

Columbus made four voyages west across the Atlantic, establishing Spanish colonies on the islands he passed and claiming the region for Spain. Right up to

THE NEW WORLD?
The lands visited by Columbus disappointed him, for he did not find the walled cities and fabulous wealth of China and Japan he expected. Yet he remained convinced that he had sailed to Asia and never realized that what he had discovered was a continent previously unknown to Europeans.

◀ NATIVE AMERICANS
The Arawak peoples of the West Indies lived off the abundant fruits and berries of the islands. They lived in shelters that they built out of palm leaves and branches. Most people did not wear anything, although some wore clothes for ceremonies.

▼ TOBACCO
While in Cuba, Columbus saw the Arawak people roll the dried leaves of the tobacco plant into a tube, set light to it and smoke it. Smoking soon became a popular pastime throughout Europe. Below you can see tobacco leaves being dried in a shed.

◀ FERDINAND AND ISABELLA
When Ferdinand of Aragon married Isabella of Castile in 1469, Spain became a united country for the first time since the Roman Empire. Isabella sponsored Columbus's first voyage.

CHRISTOPHER COLUMBUS

▲ COCONUT PALMS
During his travels, Columbus saw many crops unknown to Europeans, including coconuts, pineapples, potatoes and sweetcorn.

▼ THE *SANTA MARIA*
Columbus's flagship was the Santa Maria, a three-masted, square-rigged cargo ship capable of holding up to 40 crew. The other two ships, the Niña and Pinta, were much smaller.

his death in 1506, he remained convinced that he had sailed to Asia, although he failed to find proof. Because he had sailed west, the new islands he had come across became known as the West Indies.

Few people accepted his claims. In 1502 Amerigo Vespucci (1451–1512) returned to Europe from an expedition down the east coast of South America. He was certain that the lands were not part of Asia, but part of a continent unknown to Europeans. He called it *Mundus Novus* – the New World. In 1507 a German geographer, Martin Waldseemüller, renamed it America in honour of Amerigo Vespucci. What Columbus had actually discovered was of far greater importance than a lengthy sea route to Asia. By sailing west, he had stumbled upon the American continent. As a result, within a few years the history of both America and Europe was completely transformed.

▼ THE VOYAGES OF COLUMBUS
Over the course of four voyages, Columbus sailed round most of the Caribbean islands and explored the coasts of South and Central America, believing that he had discovered a new route to Asia.

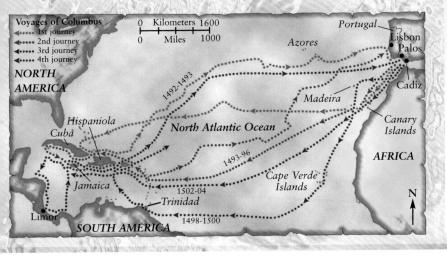

Key Dates

- 1492–93 Columbus makes his first voyage to the West Indies, finding the Bahamas, Cuba and Hispaniola.

- 1493–96 His second voyage takes him throughout the West Indies. He builds settlements on Hispaniola and explores Jamaica.

- 1498–1500 On the third voyage he sails between Trinidad and South America and is the first European to land in South America.

- 1502–04 Fourth voyage, along the coast of Central America.

Conquering the New World

In the years after the historic voyages of Columbus a wave of Spanish explorers descended on Central and South America. They were searching for treasure.

Vasco de Balboa (1475–1519) was one of these adventurers. He was a colonist living in Hispaniola (Haiti and the Dominican Republic), who fled to Central America to escape his debts. In September 1513 he set off into the interior of the country in search of gold. Twenty-seven days later he gazed westwards across a vast sea, becoming the first European to look at the eastern shore of the Pacific Ocean.

In November 1518 a second expedition left the Spanish colony of Santiago in Cuba, bound for Mexico. Previous expeditions had reported that there were vast temples and huge amounts of gold there. The 11 ships and 780 men were commanded by Hernán Cortés, a Spanish lawyer who had gone to the West Indies to seek his fortune. Cortés sailed along the coast for some months, raiding local towns and gaining valuable intelligence, then set off inland to the Aztec capital of Tenochtitlán.

Although the Aztecs were immensely skilled people, they were no match for the Spanish. The Aztecs had no gunpowder, and horses were unknown in the Americas. Cortés enlisted the help of the Aztecs' many enemies, then entered the city and captured its ruler, Montezuma. Cortés finally secured Tenochtitlán in August 1521, with only 400 men. The mighty Aztec Empire now became the province of New Spain.

Soon, rumors began to circulate about another rich empire, this time in South America. In 1530 Francisco Pizarro set out to conquer it with only 168 soldiers. The Inca Empire he found was weakened by civil war and an epidemic (probably smallpox). Once again, the Spanish soldiers overwhelmed the enemy.

By 1532 the vast Inca Empire was defeated and its huge reserves of gold and silver were now under Spanish control.

Cortés, Pizarro and the other adventurers were *conquistadores*, which

◄ KNIFE
The Aztecs were skilled craftworkers. They used wood inlaid with gems and pieces of shell and turquoise to make the handle of this sacrificial knife. This knife was given as a gift to Hernán Cortés.

◄ MONTEZUMA'S HEADDRESS
The Aztecs and Incas hunted tropical birds for their feathers. The quetzal's bright green feathers were highly prized and used in the headdress of Montezuma, the last Aztec ruler.

THE INCAS
The Incas were a hill tribe from Peru. Over the course of 300 years, they came to dominate the whole of the Andes mountains. By 1500 their empire stretched for more than 2,485 miles (4,000 km). Although they had no wheeled transport, they built a huge network of roads and large cities of stone. They seem to have had no alphabet, so could not read or write. Despite this, their civilization was as advanced as any in Europe. The Incas were overthrown by Pizarro's small army.

▲ PIZARRO
The Spaniard Francisco Pizarro (1475–1541) went to the Americas to seek his fortune. He was spectacularly successful, crushing the powerful Inca Empire.

► QUIPU
Special officials kept records of taxation, population figures and other statistics on quipus. A quipu is a series of vertical knotted strings, of varying length and color, that hang from a horizontal cord. The length and color of each string, its position and the type of knot record the information.

◄ GOLD LLAMA
Llamas were valued by the Incas for their meat, wool and as beasts of burden. Gold figurines were made to show their importance.

Conquering the New World

◀ TENOCHTITLAN
The Aztecs' capital city had a population of 200,000, bigger than any Spanish city, yet Cortés and his 400 men managed to capture it using trickery and deceit.

means conquerors in Spanish. The conquistadores were brutal and often dishonest. They went in search of wealth, and to convert everyone they met to Christianity. Their conquests stretched the length of the Americas, from Mexico to Chile. Within 50 years of the expedition by Columbus, the Americas were under European control.

▶ MACHU PICCHU
The Incas established the city of Machu Picchu in a strategic position, protected by the steep slopes of the Andes mountains. It was built of stone blocks fitted together without mortar. Temples, ceremonial places and houses made up the 143 buildings. The city was so remote that the Spanish failed to discover it, and it was forgotten until an American explorer found it in 1911. It is located in south Peru.

Key Dates

- 1100s Incas start to dominate central Peru.
- 1325 Aztecs found the city of Tenochtitlán.
- 1430 Incas begin to expand north along the Andes.
- 1450s Incas build Machu Picchu.
- 1500 Aztec and Inca empires at their greatest extent and power.
- 1513 Vasco de Balboa first sees Pacific Ocean.
- 1521 Spanish capture Tenochtitlán and take over the Aztec Empire.
- 1532 Inca Empire conquered and under Spanish control.

Round the World

▲ FERDINAND MAGELLAN
Magellan (1480–1521) was a Portuguese sailor who quarrelled with the Portuguese king and left the country in 1514 to work for the King of Spain. His round-the-world fleet sailed under the Spanish flag.

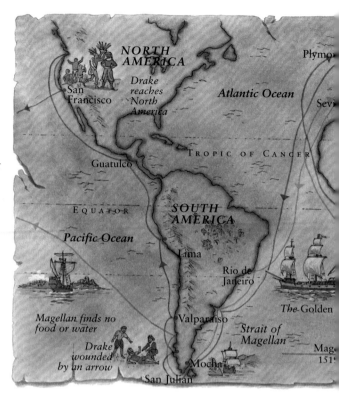

European nations were entranced by stories about the vast wealth of Asia. Travelers and merchants told of treasures in India, China, Japan and the spice-rich islands off their coasts. Throughout the 1500s sailors made epic voyages to seek out new routes to this wealth.

After the voyages of the Portuguese to the Indian Ocean and Columbus to America, Spain and Portugal made the Treaty of Tordesillas in 1494. The two countries divided the undiscovered world between them. They drew a line on a map and agreed that everything to the west of it was the property of Spain, and everything to the east belonged to Portugal. South America was cut in half by the line.

Spanish explorers still wanted to find a new route to Asia by going west, as Columbus had tried to do. Columbus had discovered America when he went west, although he thought it was Asia. His successors had to find a way around America in order to get to Asia. In 1519 Ferdinand Magellan left Spain with five ships and 260 men to find a route to the rich Spice Islands (the Moluccas, now part of Indonesia). In 1520 he sailed through the straits at the tip of South America and into the Pacific Ocean. He sailed northwest, and in 1521 reached the Philippine Islands.

Magellan never reached the Spice Islands, because he was killed in a skirmish in April 1521. But one of his ships managed to get there. The *Victoria* was captained by Juan de Elcano (1476–1526). When the crew

PRIVATEERS AND PIRATES
Treasure ships heading for Spain laden with riches were soon noticed by Spain's main enemies, France and England. In wartime, both nations allowed privately owned ships (known as privateers) to attack Spanish ships and keep the booty. However, privateers often attacked in peacetime. Illegal pirate ships also joined in. Spain considered everyone who attacked one of its ships to be a pirate.

▲ JOLLY ROGER
During the 1600s, pirate ships began to fly the Jolly Roger – a black flag with a skull and crossbones on it – to announce to other ships their hostile intent. Each pirate ship had its own flag, but the most feared was the plain red flag, which meant death to every sailor who resisted a pirate takeover.

▼ DOUBLOONS
The Spanish mined precious gold and silver in the Americas. Some was made into coins to take to Spain. Gold was minted into doubloons, silver into pieces of eight.

▶ PIRATES
Pirates faced death if they were captured. Escaped slaves and convicts often became pirates. When they were attacked by a pirate ship, sailors often joined the pirates, hoping to get rich.

◀ ROUND THE WORLD
Both Magellan and Drake sailed in a westerly direction. From Europe their voyages took them to the south Atlantic Ocean, round Cape Horn, across the Pacific and Indian oceans, then back via the Cape of Good Hope and Atlantic.

▲ THE *GOLDEN HIND*
Francis Drake's flagship, the Golden Hind, was originally called the Pelican. It had three masts and was the largest of the five ships in the fleet.

reached the Spice Islands they loaded the ship with spices and headed home across the Indian Ocean.

In trying to find a westerly route to the Spice Islands, Magellan and his sailors had inadvertently become the first people to circumnavigate the Earth. Others followed Magellan. Francis Drake (1543–96) was an English seafarer and pirate with a successful record of raiding Spanish ships. In 1577 he set off to explore the Pacific Ocean, attacking Spanish treasure ships and collecting their gold as he went. In the Spice Islands he bought about 13,226 pounds (6 tons) of valuable cloves. When he returned to England, this treasure was worth about $15.5 million in today's money.

▶ SPICES
Spices were highly valued in Europe for flavoring meat after it had been salted to preserve it, as well as for adding flavor to other foods and drinks. Cloves, nutmeg, cinnamon, pepper and other spices all grew wild in the Far East. They had also been cultivated for centuries and were sold in markets as they are today.

cloves

cinnamon

Key Dates

- 1519–21 Magellan sails from Spain to the Philippines across the Pacific Ocean.

- 1521–22 Juan de Elcano completes the first circumnavigation of the world.

- 1520s First treasure ships bring Aztec gold back to Spain.

- 1545 Silver discovered in vast Potosí mine in Bolivia. It was the world's biggest single source of silver for the next 100 years.

- 1545 Major silver mine opened in Zacatecas, Mexico.

- 1577–80 Drake is the second person to sail around the world.

Into Canada

▲ JOHN CABOT
The adventurer John Cabot (1450–99) was probably born in Genoa in Italy. He traded in spices with the Arabs before moving to England.

In about 1494 an Italian merchant called John Cabot arrived in England. Like Columbus, he planned to sail west across the Atlantic in search of the Spice Islands of eastern Asia. However, he proposed to make the voyage at a more northerly latitude, making the journey shorter. Cabot needed to find someone to finance his trip. After rejection by the kings of both Spain and Portugal, Cabot took his idea to King Henry VII of England. Henry had previously refused to sponsor Columbus. This time he was aware of the riches of the New World, and was eager to support Cabot so that he could profit from any discoveries.

In May 1497 Cabot set sail from Bristol on board the *Matthew*. A month later he landed in Newfoundland, off the east coast of Canada, which he claimed for England. He had not found Asia, nor had he found wealth, but he had discovered rich fishing grounds and lands not yet claimed by Spain.

The French set out to explore these new lands. In 1534 Jacques Cartier (1491–1557) sailed from St Malo. Like Cabot, he too was searching for a new, northerly route to Asia. He sailed round the mouth of the great St Lawrence River, returning the following year to sail up it to present-day Montreal. He struck up good relations with the Huron Indians who lived there, who told him about the riches of the kingdom of Saguenay, further west up the St Lawrence. In 1541 Cartier decided to return to find Saguenay. But not surprisingly he failed to do so, because Saguenay was an imaginary place. The Hurons had made up the story about this marvellous kingdom, full of treasures, to please their French visitors!

◀ MONTREAL
When Cartier sailed up the St Lawrence River in 1535, he got as far as the wooden-walled Huron village of Hochelaga. Cartier climbed the hill behind it, naming it Mont Réal (Mount Royal), the present-day Montreal.

NATIVE AMERICANS
Numerous tribes of Native Americans lived in the woods and plains of the St Lawrence valley. Five of the main tribes – the Mohawk, Onondaga, Seneca, Oneida, and Cayuga – joined together to form the Iroquois League in the early 1600s to protect themselves from other powerful tribes in the area.

◀ A HURON BRAVE
The Hurons welcomed the French to North America, trading furs and other goods with them and acting as guides and advisers. They also enlisted the French to help fight their wars with the Iroquois, who were their deadly enemies.

▼ FUR TRADE
The rivers and woods of Canada teemed with wildlife, providing furs for clothing and meat for food. Animal pelts, particularly from the seal, otter and beaver, were prized by the Europeans. They traded guns and other goods to obtain the skins from the Native Americans.

▲ A SCALP
Fierce warfare between the different tribes was common. The most important trophy a brave could win in battle was the scalp of his opponent. Skin and hair were removed in one piece and then displayed on a wooden frame.

INTO CANADA 31

▶ QUEBEC
When de Champlain visited Canada in 1608, he built a wooden fort on a hill overlooking the St Lawrence River at a point where it narrowed considerably. The Native Americans called the place Kebec, and today it is known as Quebec.

Fort built of wood

Balcony for strategic lookout

Cannon positioned for quick firing

Bridge for crossing the St Lawrence

Fur traders and fishermen followed Cartier's route up the St Lawrence. But it was not until the next century that the French abandoned their search for a new route to Asia and began to settle in Canada. Samuel de Champlain (1567–1635) explored the east coast of North America and went inland as far as the Great Lakes. In 1608 he founded the city of Quebec, the first permanent French settlement in North America. The continent was now open for European colonization.

▼ EXPLORING CANADA
After John Cabot's exploratory voyage in 1497, the Frenchmen Jacques Cartier and Samuel de Champlain explored the valley of the St Lawrence River and claimed the region for France. De Champlain founded the city of Quebec.

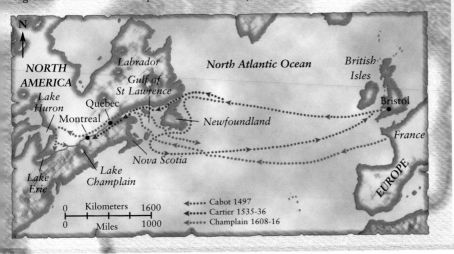

Key Dates

- 1497 John Cabot claims Newfoundland for England.
- 1534 Jacques Cartier explores St Lawrence estuary in Canada.
- 1535–36 Cartier sails up St Lawrence as far as Montreal.
- 1603 De Champlain sails up the St Lawrence to Montreal.
- 1604 De Champlain explores from Nova Scotia to Cape Cod.
- 1608–09 De Champlain founds settlement of Quebec.
- 1615 De Champlain explores Lakes Huron and Ontario.

The Northwest Passage

▲ POLAR BEARS
Polar bears were a constant threat to Arctic explorers, although their meat was a useful supplement to rations.

When Spain and Portugal divided up the undiscovered world between them in 1494, other European nations were prevented from sailing to Asia round the south of Africa or America. The Pope had split the world between Spain and Portugal, whose ships stopped British and Dutch traders from sailing south across the Atlantic. The only way left for the British and Dutch was to sail round the top of the world. For more than 300 years, explorers tried to find a route through the Arctic Ocean, either around Canada, or around Siberia. Their efforts, however, proved fruitless.

In 1576 Queen Elizabeth I of England sent Martin Frobisher off to find a northwest passage to China. He reached Baffin Island, then returned home with rocks of gold. These turned out to be iron pyrites, or "fool's gold," and had no value.

The Englishman Henry Hudson was an experienced navigator who discovered a big river on the east coast of America in 1609. It was named the Hudson River after him. But he could not find a northwest passage either. In 1610 he sailed his ship *Discovery* around northern Canada before heading south towards what he hoped would be the Pacific Ocean. In fact it turned out to be the vast but landlocked bay now called Hudson Bay. His crew refused to continue and mutinied, setting Hudson and the loyal members of his crew adrift in an open boat.

Over the next two centuries, a number of expeditions mapped the north coast of Canada but failed to make much headway through the maze of Arctic islands. Interest in the project faded. Then in 1817 the British government offered a large reward to whoever could find a northwest passage. Many explorers set out but failed to find it. In 1844 the British Royal Navy organized a big expedition led by John Franklin. But he fell victim to the extreme cold and died in 1847, along with all his men.

▲ THE ARCTIC
The seas to the north of Canada are filled with islands. Between them are narrow channels which freeze solid every winter, and are full of floating pack ice and icebergs during the short summer.

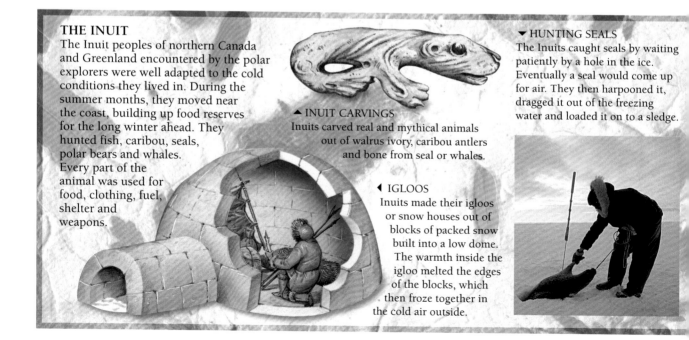

THE INUIT
The Inuit peoples of northern Canada and Greenland encountered by the polar explorers were well adapted to the cold conditions they lived in. During the summer months, they moved near the coast, building up food reserves for the long winter ahead. They hunted fish, caribou, seals, polar bears and whales. Every part of the animal was used for food, clothing, fuel, shelter and weapons.

▲ INUIT CARVINGS
Inuits carved real and mythical animals out of walrus ivory, caribou antlers and bone from seal or whales.

◀ IGLOOS
Inuits made their igloos or snow houses out of blocks of packed snow built into a low dome. The warmth inside the igloo melted the edges of the blocks, which then froze together in the cold air outside.

▼ HUNTING SEALS
The Inuits caught seals by waiting patiently by a hole in the ice. Eventually a seal would come up for air. They then harpooned it, dragged it out of the freezing water and loaded it on to a sledge.

The Northwest Passage 33

◀ MUTINY
In June 1611 the crew of Hudson's ship Discovery mutinied in Hudson Bay. They put Hudson, his son and his loyal crew in an open boat with no oars. They were left to die.

Over the next decade, more than 40 expeditions went to look for Franklin. In 1859 the last message he left was found on King William Island. Some of those ships found the Northwest Passage, although none sailed through it. It was not until 1906 that the Norwegian polar explorer Roald Amundsen sailed from east to west along the north coast of Canada to the Pacific Ocean. By then, Spain and Portugal no longer controlled the southern seas, and the Panama and Suez canals were open for international shipping. The Northwest Passage was now just a long stretch of icy, treacherous water and of no commercial or political interest whatsoever.

▼ THE NORTHWEST PASSAGE
It took more than 300 years for explorers to navigate the Northwest Passage, and many died in the attempt. Finally, Roald Amundsen succeeded in 1906. By that time, there were easier ways of getting to Asia.

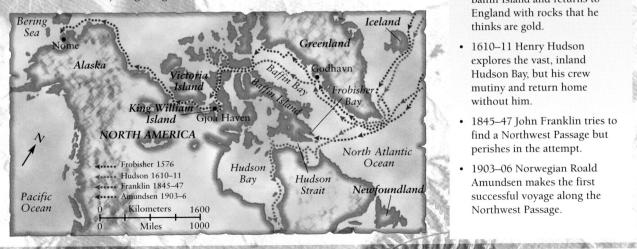

Key Dates

- 1576 Martin Frobisher lands on Baffin Island and returns to England with rocks that he thinks are gold.
- 1610–11 Henry Hudson explores the vast, inland Hudson Bay, but his crew mutiny and return home without him.
- 1845–47 John Franklin tries to find a Northwest Passage but perishes in the attempt.
- 1903–06 Norwegian Roald Amundsen makes the first successful voyage along the Northwest Passage.

The Northeast Passage

▲ SEALS
Siberian peoples, and explorers seeking the Northeast Passage, hunted seals for food. Their skins also had many uses.

While English sailors concentrated on finding a northwest passage round the north of Canada, the Dutch were seeking a northeast passage to the north of Russia and Siberia. The Dutch were good sailors and their fishing and whaling fleets regularly sailed in the Arctic Ocean, but even so they were unsure whether a northeast passage really did exist.

The Dutch followed in the wake of the Englishman Hugh Willoughby (1510–54), who had succeeded in sailing as far as the large island of Novaya Zemlya. But on his return journey, he perished in the pack ice off Murmansk on the Kola Peninsula in 1554.

In 1594 the Dutch mariner Willem Barents (1550–97) set sail on the first of his expeditions to find the Northeast Passage. He too was unsuccessful, although on his third voyage in 1596 he discovered Bear Island, which got its name after his crew had a fight with a polar bear. He also found the rich fishing grounds of the Spitsbergen archipelago, which was to become hugely profitable for Dutch hunters of whales, seals and walruses. Then Barents's ship became trapped and damaged by the winter pack ice. He set off to row and sail the 1,590 miles (2,560 km) to Kola, but died of starvation at sea. His crew

▲ SIBERIA
The northern coast of Siberia lies inside the Arctic Circle. Here the temperature barely reaches above freezing point in summer and drops far below it in winter. Little grows in such an inhospitable landscape, although animals such as reindeer live there.

survived and managed to return home once the ship was free.

After the failure of Barents's trip, there were no more expeditions until the Russian explorer Semyon Dezhnev (1605–72) sailed round the eastern tip of Siberia into the Pacific Ocean, proving that Asia and America were not joined. This knowledge did not reach Europe for many years, and the Arctic Ocean was mainly left to the Siberians to fish. It was not until the late 1800s that

THE ARCTIC OCEAN
Although the waters north of Siberia do not freeze over as much as those north of Canada, the Arctic Ocean is still a harsh place. The ice-free summer months are short and ships risk being caught and trapped in pack ice during the winter, which can last for up to nine months.

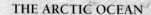

◀ THE *VEGA*
The 661,387-pound (300 ton) *Vega* was built in Germany as a whaling ship. It was constructed of oak with an outer skin of tougher wood to protect it against the ice. The ship had sails and a powerful steam engine. Nils Nordenskjöld made the first successful transit of the Northeast Passage in it during 1878–79.

◀ ICE-BREAKER
Today the Northeast Passage is kept reasonably free of ice by a fleet of ice-breakers. These specially strengthened ships clear a passage to allow shipping through the pack ice.

▶ WHALES
The first people to explore the Arctic Ocean were whalers (whale hunters) from ports in northern Europe. They sailed the ocean in reinforced ships. Whales were caught for their meat, blubber and bone.

THE NORTHEAST PASSAGE 35

◀ AN ARCTIC SHELTER
In the winter of 1596 the ship of Barents and his crew of 20 men was trapped by ice in the Arctic. The men survived by building a hut out of driftwood. The hut measured 33 x 20 feet (10 x 6 m) and contained a fireplace, with a chimney to escape through if the hut was buried by snow. It even had a primitive Turkish bath made out of a barrel.

there was interest in the Northeast Passage again, when Russia and other European countries realized that the great rivers and forests of Siberia might be a rich hunting ground. In 1878 the Finnish polar explorer Nils Nordenskjöld (1832–1901) set out from southern Sweden on board his ship, the *Vega*. Keeping close to the Siberian coast, he voyaged east until ice near the Bering Strait blocked his way. The following July he sailed into the Pacific Ocean. The quest was over and the Northeast Passage was now open for commerce.

▲ THE NORTHEAST PASSAGE
For more than 300 years, explorers sailed northeast from Europe in the hope of finding a way along the top of Siberia and around into the Pacific Ocean. They were looking for an easy route to the riches of Asia.

Key Dates

- 1554 Hugh Willoughby reaches Novaya Zemlya.
- 1594 Barents sails into Kara Sea east of Novaya Zemlya.
- 1595 Barents's second voyage ends in failure in the Kara Sea.
- 1596 Barents discovers Bear Island and the rich fishing grounds of Spitsbergen.
- 1597 Barents dies on the sea that now bears his name.
- 1648 Semyon Dezhnev sails round eastern tip of Siberia.
- 1878–79 Nils Nordenskjöld navigates the Northeast Passage.

Exploring Asia

▲ PETER THE GREAT
Peter I (Peter the Great) was Tsar of Russia from 1682 to 1725. Under him, the backward country was transformed into a great European power.

At the start of the 1700s Russia had a dynamic ruler, Tsar Peter I. He built up a big navy and army, reorganized the government and constructed the country's new capital of St Petersburg. In previous years Russia had extended its territory right across Siberia to the shores of the Pacific Ocean. However, few Russians had any idea what their new land contained, or whether it was joined to America, so Peter the Great decided to find out.

Vitus Bering (1681–1741) was born in Denmark. He was a superb administrator, and Peter invited him to help modernize the Russian navy. In 1724 Peter appointed him to lead a large expedition across Siberia. The expedition left St Petersburg in 1725 and reached the Pacific Ocean two years later. There Bering and his men built a ship, the *St Gabriel*, and sailed up the coast and into the Arctic Ocean. From what he saw, Bering was satisfied that Siberia and America were not linked and returned to St Petersburg in 1730. In 1732 Bering was put in charge of a huge new undertaking. The Great Northern Expedition consisted of more than 3,000 men, including 30 scientists and 5 surveyors, with 13 ships and 9 wagonloads of scientific instruments. Its task was to explore the entire northern coast of Siberia as well as the seas to its east.

Over the next ten years five teams mapped the northern coast and the great rivers that flowed north through the country towards it. Bering concentrated on the seas beyond Siberia. This time he sailed across the Pacific to Alaska, returning to the Kamchatka Peninsula along the string of islands called the Aleutian Islands.

Bering died in 1741, before the expedition was finished, but he achieved a great deal. His team had

▲ CROSSING SIBERIA
Travelers in Siberia used teams of trained reindeer or huskies to pull sledges bearing food and other provisions. People sometimes wore wide snowshoes to stop themselves sinking into the snow.

EASTERN ASIA
During the 1500s, Europeans began to travel to China and Japan. Most were Jesuit missionaries, who were trying to convert people to Christianity. However, eastern Asia was still mainly closed to foreigners, and little was discovered about these strange and distant lands.

◀ JESUIT PRIEST
The Jesuits were a Roman Catholic religious order formed in 1534 by Ignatius Loyola. Their main aim was to convert Muslims to Christianity, but they soon expanded their work, opening missions in India and China.

▲ FRANCIS XAVIER
Francis Xavier (1506–52) was a Spanish Jesuit who travelled round India before visiting Japan in 1549. He admired Japanese people for their sense of honor, and made many converts.

▶ PRAYER WHEEL
Siddhartha Gautama was an Indian prince who became known as the Buddha. The religion of Buddhism is based on his teachings. From about 400 BCE it began to spread throughout eastern Asia. Many Tibetan Buddhists use a prayer wheel for saying their prayers.

Exploring Asia 37

▲ HUNTERS' PREY
The Siberian tiger lives in southeast Siberia, near the border with China. Its pelt (skin) was much prized by fur trappers.

▶ DEATH OF BERING
In 1741 Bering started the voyage back from the Aleutian Islands off Alaska. He reached an island near the Kamchatka Peninsula, where he died of scurvy and exposure. The island is now named after him.

mapped Siberia and opened up both Siberia and Alaska to Russian fur traders. By 1800 Alaska was part of the Russian Empire. Although Semyon Dezhnev had discovered a century earlier that Siberia and America were separated by sea, he left no records and few people were aware of his work. Bering confirmed these findings, so in his honor, the strait between the two continents is named the Bering Strait.

▶ LHASA
In Tibet, the isolated city of Lhasa was the center of Tibetan Buddhism. In 1658 the German Jesuit John Grueber (1623–80) and the Belgian Albert d'Orville (1621–62) set out from China to find an overland route to India so as to avoid hostile Dutch ships on the sea route. In 1661 they entered Lhasa – the first Europeans to set eyes on the mystical city with its palaces and great temple complexes.

Key Dates

- 1549 Francis Xavier in Japan.
- 1661 Grueber and d'Orville visit Lhasa in Tibet.
- 1725–29 Bering crosses Siberia and explores the sea between Siberia and Alaska.
- 1732 Bering organizes Great Northern Expedition.
- 1734–41 Bering crosses Siberia and explores coast of Alaska.
- 1734–42 Five teams of explorers map the northern Siberian coast and the Ob, Yenisei and Lena rivers.

Advancing into America

▲ JEFFERSON
In 1803 President Thomas Jefferson bought the Louisiana Territory from France, more than doubling the size of the United States.

Two hundred years after Columbus landed in the West Indies, Europeans still knew surprisingly little about the enormous American continent to the north. The Spanish explored Florida and the Gulf of Mexico, the English established colonies on the east coast, and the French sailed up the St Lawrence River and settled in Canada. But the vast lands that lay in between remained a mystery.

In 1541 the Spaniard Hernando de Soto set out to explore Florida and became the first European to set eyes on the wide southern reaches of the Mississippi River. Unfortunately he died soon afterwards and the Spanish failed to explore further. More than a century later, hundreds of miles to the north, Louis Jolliet (1645–1700) and the French Jesuit missionary Father Jacques Marquette (1637–75) discovered a route to the Mississippi River from the Great Lakes. They explored the river as far south as Arkansas. However, it was another Frenchman, Robert de la Salle (1643–87), who became the first European to sail down the river to its mouth on the Gulf of Mexico. He claimed the land in this area for his country, naming it Louisiana after the French king, Louis XIV.

▶ SAGAJAWEA
In 1804 Lewis and Clark were joined by Sagajawea, a member of the Shoshone tribe. She spoke many native languages and acted as the interpreter on the expedition.

Over the next century, European influence in North America changed considerably. The Spanish still controlled Mexico and Florida, but the English had thrown the French out of Canada. Most importantly, the English colonists rebelled against their own country and set up an independent United States that stretched from the Atlantic coast to the east side of the Mississippi River. The west side – Louisiana Territory – still belonged to France, but in 1803 the French sold it to the United States.

US President Thomas Jefferson wanted to find out more about this vast new land he had bought. In 1804 he sent two men to explore it. They were his personal secretary, Meriwether Lewis (1774–1809), and William Clark (1770–1838), a former army officer. Over the course of two years, they followed the river from St Louis up the Missouri River, over the Rockies, and

▲ CABEZA DE VACA
Alvar Núñez Cabeza de Vaca (1490–1556) sailed with de Narváez around the Gulf of Mexico. The fleet was wrecked off Texas in November 1528, but Cabeza de Vaca was saved by Yaqui tribesmen. He stayed with them for five years, then set out on foot through Texas and across the Rio Grande into Mexico, reaching the safety of Mexico City in 1536.

THE NEW CONTINENT
The Spanish were the first Europeans to explore North America moving northwards from their empire in Mexico. Pánfilo de Narváez (1470–1528) explored the Gulf of Mexico, while Hernando de Soto (1500–42) became the first European to see the Mississippi River in 1541. These were the first of many people to push across this huge new continent.

▶ THE MISSISSIPPI RIVER
The mighty Mississippi River flows south across North America to the Gulf of Mexico. The discovery of its northern reaches by Jolliet and Marquette opened up America to European explorers and settlers.

◀ BISON
For 350 years European settlers hunted the bison herds of the plains almost to extinction, wiping out the Native Americans' main source of food and clothing.

down the Columbia River to the Pacific coast, before returning via the Yellowstone River to St Louis.

The success of the expedition convinced the US government that the Louisiana Territory was suitable for people to live in. Within a generation, settlers were pouring across the Mississippi to start a new life on the Great Plains and the Pacific coast. The expansion of the United States across the American continent had begun.

▲ SHOOTING RAPIDS
Lewis and Clark used canoes to navigate the dangerous Missouri, Columbia and Yellowstone rivers.

▶ GRIZZLY BEARS
Bears were a menace to the expedition. One chased six men from Lewis and Clark's party into the Missouri River.

▲ SPREADING OUT ACROSS AMERICA
Robert de la Salle's two voyages around the Great Lakes and down the Mississippi River, and the expedition of Lewis and Clark up the Missouri River, did much to open up North America to traders and eventually to settlers.

Key Dates

- 1527–28 De Narváez explores the Gulf of Mexico.
- 1528–36 De Vaca explores Texas.
- 1541 De Soto is first European to see the Mississippi.
- 1672 Marquette and Jolliet explore the upper Mississippi.
- 1678–80 La Salle explores the Great Lakes.
- 1680–82 La Salle sails down Mississippi to Gulf of Mexico and claims the region for France.
- 1804–6 Lewis and Clark explore the Missouri River and routes to the Pacific.

Across the Pacific

▲ ABEL TASMAN
The two voyages of Abel Tasman did much to map the uncharted lands of the southern oceans.

Ever since the time of the ancient Greeks, people in Europe had imagined that there was a great continent lying on the other side of the world. They reasoned that since there was a Eurasian continent in the northern hemisphere, there must be a similarly large continent in the southern hemisphere, in order to balance the world! The only problem was that no one had ever managed to discover where this southern continent actually was.

Several sailors employed by a trading organization, the Dutch East India Company, stumbled across unknown land during their voyages. In 1605 Willem Jansz (1570–1629) sailed south from New Guinea and found the northern tip of Australia. In 1615 Dirk Hartog (1580–1630), travelling to Indonesia, sailed too far east and landed in Western Australia. Both sailors reported that this new land was too poor to bother with. So the Dutch East India Company took no further action, as it was interested in trade, not exploration.

In 1642 the company changed its mind and began to search for Terra Australis Incognita, or "unknown southern land." In 1642–43 Abel Tasman (1603–59) sailed around the Indian and Pacific Oceans in a huge

▲ TASMANIA
When Tasman landed on a new island in November 1642, he named it Van Diemen's Land, after the governor-general of Batavia in the East Indies. It was later renamed Tasmania, after Tasman.

circle without discovering a southern continent, although he did find the island later named after him – Tasmania – and New Zealand. In 1643–44 he explored the Australian coastline that Jansz and Hartog had found. Tasman thought that the land to the south of New Guinea was not part of a southern continent, but

THE SOUTH PACIFIC
Although both Magellan and Drake crossed the Pacific, their routes took them north of the many island groups. During the following centuries these islands were gradually reached by Europeans: Alvaro de Mendaña (1541–95) reached Tuvalu and the Solomon Islands; Pedro Quirós (1565–1614) got to Vanuatu; and Tasman saw Fiji and Tonga. Louis Bougainville (1729–1811) mapped the region, but did not reach eastern Australia because the Great Barrier Reef was in the way.

◀ THE GREAT BARRIER REEF
This 1,243 mile-long (2,000 km) coral reef runs along the coast of northeast Australia. It is made up of the skeletons of millions of tiny sea creatures and is home to lots of marine life. It prevented Bougainville and other explorers from landing in Australia.

◀ THE SOLOMON ISLANDS
The first South Pacific islands to be explored by Europeans were the Solomon Islands, off the coast of New Guinea, which were found by de Mendaña in 1568. Over the next 200 years the rest of the islands in the region were slowly explored and mapped by visiting Europeans.

▶ BOUGAINVILLEA
On his round-the-world voyage Louis Bougainville took a botanist with him. One of the plants they brought back to Europe was a flowering climber now named bougainvillea in his honor.

Across the Pacific

▶ **DUTCH EAST INDIA COMPANY**
In 1602 the Dutch set up a company to coordinate their trading activities in the East Indies. They established trading posts, like the one pictured here, in India, China and Japan, soon controlling the local spice trade.

he did not find out whether it was connected to New Guinea or whether it was an island.

Strangely enough, Luis Torres (c1570-1613) had already proved that New Guinea was an island. In 1607 he sailed right around New Guinea through the strait which now bears his name, showing that it was an island. Therefore the land to its south, Australia, could not be attached to it. However, Tasman did not realize the significance of this discovery, so the mystery concerning the great southern continent, and the unnamed land that lay above it, remained unsolved.

▼ **TASMAN AND BOUGAINVILLE**
Neither Abel Tasman nor Louis Bougainville actually landed in Australia, although both did much to increase knowledge of the continent. Bougainville's voyage round the world established a French presence in the South Pacific.

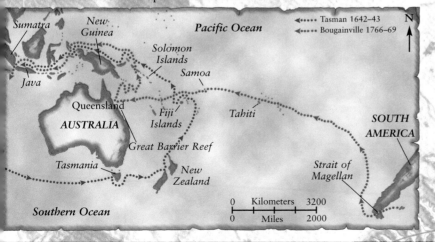

Key Dates

- 1567–69 De Mendaña discovers the Solomon Islands.
- 1602 Dutch East India Company is established.
- 1605 Willem Jansz explores Queensland.
- 1615 Dirk Hartog discovers Western Australia.
- 1642–43 Tasman discovers Tasmania and sees New Zealand and the Fiji Islands.
- 1643–44 Tasman maps north coast of Australia.
- 1766–69 Louis Bougainville circumnavigates the world.

Captain Cook

▲ CAPTAIN COOK
James Cook (1728–79) sailed in the merchant navy for ten years and became an experienced navigator and seaman before he joined the British Royal Navy in 1755.

By the 18th century Europeans were still not sure of the shape and size of the strange land – *Terra Australis Incognita* – in the southern hemisphere. They did not even know for sure whether this mysterious new continent actually existed. By now, the British had overtaken the Dutch as the major trading nation in the world, and their Royal Navy ruled the waves. In 1768 the British Navy sent an expedition to the South Seas to search for the southern continent. James Cook was the ideal choice to lead it – he was an expert in navigation and an experienced seaman, having spent more than ten years on merchant ships.

Cook set sail from Plymouth, England, in August 1768. In April 1769 he reached Tahiti, where he and his crew were impressed by the warm climate and the beautiful plants and wildlife. Then he sailed southwest to New Zealand, the west coast of which had been discovered by Tasman. By steering a course in the shape of a figure eight, he found that New Zealand was two islands, not one. Cook continued west, landing at a place that is now called Botany Bay, in Australia, which he claimed for Britain. He then sailed up the coast until he got to the Great Barrier Reef, where the *Endeavour* ran aground and had to be repaired. He then went through the Torres Strait and home to England across the Indian and Atlantic oceans. Cook made two further voyages to the South Seas. His second trip, in 1772–75, took him towards the South Pole, which was where Cook thought the southern continent lay. His final voyage, in 1776–79, took him north in a search for an inlet into the Arctic Ocean.

▼ BAY OF ISLANDS, NEW ZEALAND
Cook visited many fine harbors on his voyages to New Zealand. The Bay of Islands, shown below, is on New Zealand's North Island.

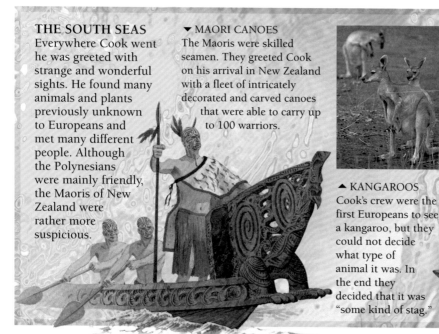

THE SOUTH SEAS
Everywhere Cook went he was greeted with strange and wonderful sights. He found many animals and plants previously unknown to Europeans and met many different people. Although the Polynesians were mainly friendly, the Maoris of New Zealand were rather more suspicious.

▼ MAORI CANOES
The Maoris were skilled seamen. They greeted Cook on his arrival in New Zealand with a fleet of intricately decorated and carved canoes that were able to carry up to 100 warriors.

▲ KANGAROOS
Cook's crew were the first Europeans to see a kangaroo, but they could not decide what type of animal it was. In the end they decided that it was "some kind of stag."

▶ LIME
During long voyages, most sailors developed scurvy, a disease caused by a lack of vitamin C in their diet. Cook solved this problem by feeding the crew with vitamin C-rich pickled cabbage, vegetables and limes.

◀ HONEYSUCKLE
Sydney Parkinson was an illustrator who went on the voyage. He drew many of the exotic plants that he saw on the journey. One was a type of honeysuckle.

Captain Cook 43

Cook met a tragic end when he was killed in Hawaii after a scuffle broke out on the beach. But in his three voyages, Cook finally proved that Australia and New Zealand were separate islands and not part of a larger southern continent. When a landmass was later discovered around the South Pole, Antarctica was identified as the true Terra Australis Incognita. The vast amount of scientific, botanical and navigational information Cook brought back from the South Seas was equally important. As a result of his work, exploration turned from adventure into scientific discovery.

Artist records plant information

Sorting animal skins and specimens

▶ THE ENDEAVOUR
When he was in the merchant navy, Cook sailed colliers, or coal ships, out of his home port of Whitby, Yorkshire. He therefore chose a converted collier, the Endeavour, to sail round the world in. The ship was slow but tough and spacious, with enough room for the 94 crew members and their supplies.

▼ COOK'S VOYAGES
Over the course of three voyages, between 1768 and 1779, Cook explored much of the Pacific Ocean, including eastern Australia, which he named New South Wales. He also discovered many islands, including Hawaii, where he was eventually murdered.

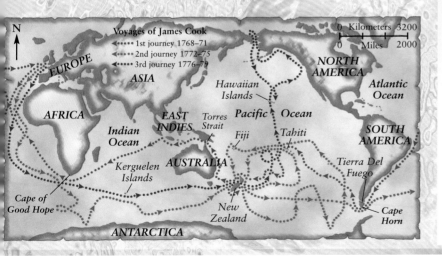

Key Dates

- **1755** Cook joins the Royal Navy and rises in rank to captain.

- **1768–71** On his first voyage, Cook sails around New Zealand and explores the east coast of Australia, claiming it for Britain.

- **1769** Cook discovers the island of Tahiti.

- **1772–75** Cook's second voyage takes him south towards Antarctica.

- **1776–79** His third voyage heads north into the Arctic Ocean, and he discovers Hawaii on the way.

- **1779** Cook is killed in a violent skirmish on a beach in Hawaii.

Trekking across Australia

▲ ABORIGINAL ART
Aborigines believe that their Ancestral Beings shaped the land and created life in a period known as Dreamtime. These Beings live on in spirit form and are represented through paintings at sacred sites, such as caves and rocks.

After Cook landed in Botany Bay, Europeans began to settle in Australia. But ninety years later they still knew little about their new country. The first settlers were convicts sent out from Britain to serve their prison sentences in Fort Jackson, now the city of Sydney. They were soon joined by farmers looking for a new start in a foreign land. There was plenty of land for everyone, so few ventured far inland from the coast.

Some intrepid explorers did investigate further, following the coastline or river valleys. In 1828 Charles Sturt (1795–1869) discovered the Darling River and then followed the Murray River to the sea. In 1844 he headed up the Murray inland. In 1840–41 Edward Eyre (1815–1901) walked along the southern coast from the town of Adelaide to find a route to the Western Australian settlement of Albany. Even by the late 1850s, the settlers still did not know what lay in the interior of their enormous country. Some thought there was a huge inland sea, while others feared it was nothing but desert. In 1859 the South Australian government offered a prize to the first person who crossed the continent from south to north.

Two expeditions set out to claim the prize. The first was led by Robert O'Hara Burke (1820–61), who was more of an adventurer than an explorer, and his young companion William Wills (1834–61). This was the biggest and most expensive expedition

▶ CAMELS
Camels were imported from India for Burke and Wills's expedition. They proved unsuitable and most eventually ended up as food for the explorers. Descendants of those that survived still live in the outback.

EARLY AUSTRALIA
The Aborigines, or Native Australians, arrived on the continent more than 40,000 years ago. They lived in isolation from the rest of the world, existing by hunting and gathering their food, catching kangaroos and other animals and harvesting wild plants, nuts and berries to eat. They were pushed off their native lands when the Europeans started colonizing Australia.

◀ MODERN ABORIGINES
After the arrival of Europeans in 1788, Aborigines were reduced to second-class citizens in their own country. About 250,000 Aborigines live in Australia today.

▼ ULURU
The name Uluru means "great pebble." It is a vast sandstone rock in central Australia which is more than 1.5 miles (2.4 km) long and is sacred to the local Aranda Aborigines. It is also known as Ayers Rock.

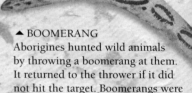

▲ BOOMERANG
Aborigines hunted wild animals by throwing a boomerang at them. It returned to the thrower if it did not hit the target. Boomerangs were often patterned like this one.

▶ INTO AUSTRALIA

Despite their extensive knowledge of the coastline of Australia, few of the early settlers knew what lay inland. Over the course of 30 years a number of expeditions set out to explore and map the interior. By 1862 the continent had been successfully crossed from south to north by Burke and Wills, although they died on the way back.

...ever organized in Australia. It consisted of 15 men, accompanied by camels and horses. It went north from Melbourne to the Gulf of Carpentaria. But the expedition was badly organized and both Burke and Wills died on the long journey back south.

John Stuart (1815–66) was more successful. He was an experienced explorer who knew how to survive in the outback. He set out from Adelaide to try and cross the continent, but was turned back by Aborigines. He set out again but was blocked by long stretches of thorny bushes. Finally, in July 1862, he succeeded in reaching Darwin. Stuart proved that the interior of Australia was indeed desert, but his journey opened up the interior for settlement and farming.

▲ PROSPECTING
The discovery of gold in Australia in 1851 brought a rush of prospectors from Europe and America, but few people struck it rich.

▼ NED KELLY
Many bushrangers, or outlaws, lived in the Australian outback. The most famous of these was Ned Kelly (1855–80), whose gang of robbers killed three policemen and robbed several banks before Kelly was finally caught and hanged in Melbourne. Kelly, who wore a tin hat to protect himself, soon became a national hero for many people.

Key Dates

- 1770 Cook lands in Botany Bay.
- 1788 First convicts to Australia.
- 1828–30 Charles Sturt crosses the Blue Mountains and reaches the Darling River.
- 1840–41 Edward Eyre walks along the south Australian coast from Adelaide to Albany in Western Australia.
- 1844–45 Sturt travels into central Australia.
- 1860–61 Burke and Wills cross Australia from south to north.
- 1862 On his third attempt, John Stuart crosses Australia from Adelaide to Darwin.

The Amazon

During the 1700s a new type of explorer emerged. While most explorers set out to make their fortune, either by finding gold or by opening up new and profitable trade routes, this new breed of explorer wanted to expand the scope of scientific knowledge.

This was a period of great scientific and intellectual debate across Europe. Scientists such as Galileo and Newton had already worked out the laws of the natural world – the movement of the planets and how motion and gravity worked. Now philosophers began to challenge existing religious beliefs with `the power of human reason, or rationality. In France a group of intellectuals compiled a 35-volume *Encyclopédie* of all knowledge. This new thinking was called "The Enlightenment." It influenced explorers, who now searched for knowledge, not for gold or glory.

South America had barely been explored since the Spanish and Portuguese conquered it in the 1500s. Two hundred years later scientists started to examine this rich and varied continent. In 1735 the French mathematician Charles-Marie de la Condamine (1701–74) went to Ecuador to record the shape and size of the Earth – the science of geodesy – by calculating its width at the equator. He was so enthralled by the wildlife there that he stayed for another ten years.

At the end of the century, the German naturalist Alexander von Humboldt (1769–1859) and the French naturalist Aimé Bonpland (1773–1858) trekked up the River Orinoco and along the Andes mountains to study plant life. Over the course of five years they recorded more than 3,000 previously unknown plant species and gathered many samples. Fifty years later two pioneering English naturalists, Henry Bates (1825–92) and Alfred Wallace (1823–1913), ventured into the Amazon rainforest. When Bates returned to England in 1859, he took with him more than 14,000 insect

▲ JAGUAR
Alfred Wallace met a jaguar, a tree-climbing big cat, when he was exploring the River Orinoco.

▶ UNKNOWN TRIBES
Explorers of the Amazon jungle discovered many tribes of people unknown to Europeans. But the arrival of settlers, and exposure to their diseases, soon killed off many of these native South American tribes.

THE AMAZON RAINFOREST
Even now, scientists have no idea how many different plants and animals live in the Amazon rainforest, as new species are constantly being discovered. Two hundred years ago the first European travelers were amazed at the sheer variety of wildlife they found, and fascinated by the tribes of people they met living deep in the jungle.

▲ ALEXANDER VON HUMBOLDT
German naturalist Alexander von Humboldt travelled extensively throughout South America, examining plants and wildlife as well as the landscape and climate. The cold sea current that flows up the west coast of South America is named in his honor.

▲ RECORDING NATURE
In the days before photography, explorers recorded what they saw by drawing it. In his sketchbooks, Henry Bates drew hundreds of the butterflies and other insects he saw on his travels.

◀ CINCHONA
Among the many new plants discovered by Aimé Bonpland was the cinchona tree. Its bark was used to make quinine, a natural cure for malaria, one of the most deadly tropical diseases.

and other specimens. Another scientist, Richard Spruce (1817–93), went back to England in 1864 with more than 30,000 plant specimens.

As a result of this scientific activity people became far more aware of the variety of life on Earth. New animals and plants were discovered, and medicines developed from some of the plants. The age of the scientific explorer was now well under way.

▶ TOUCAN ATTACK
Although toucans are normally shy and nervous, they can be aggressive. When naturalist Henry Bates attempted to capture one, he was attacked by a flock of its fellow birds.

▲ RIVER AMAZON
The Amazon in South America is the second-longest river in the world and runs east from the Andes mountains, through Brazil to the Atlantic Ocean. It flows through the world's largest rainforest, which is home to many exotic animals and plants.

◀ COLLECTING RUBBER
Rubber is made from latex, a sticky white liquid drained from the trunk of the rubber tree and collected in pots. Columbus saw locals playing with a rubber ball, but la Condamine was the first European to take rubber back home, in 1744.

▲ TREE FROG
Tree frogs were among the many new and exciting species that European naturalists encountered for the first time in the rainforest.

Key Dates

- 1735–44 La Condamine studies the shape of the Earth at the Equator. He stays there to watch the wildlife.

- 1799–1804 Alexander von Humboldt and Aimé Bonpland study botany along the River Orinoco and in the Andes.

- 1848–59 Amazon explored by Alfred Wallace and Henry Bates.

- 1849–64 Richard Spruce travels up the Amazon, collecting 30,000 plant specimens.

- 1852 Wallace returns to England, but all his specimens are lost in a fire on board ship.

Deep inside Africa

▲ RICHARD BURTON
Richard Burton was a fearless explorer. In 1853 he dressed up as an Arab and visited the holy city of Mecca, in Saudi Arabia, which was closed to non-Muslims.

The only part of Africa known to Europeans was its coastline, which for most of its length was an inhospitable place. There were few natural harbors and much of the coast was either dry desert or wet jungle. Many of the rivers flowed out into the sea through swampy deltas. As a result, the interior of Africa was too difficult for European travelers to get into.

During the late 1700s Europeans began to venture inland, exploring both the major rivers and the vast Sahara Desert. In 1770 James Bruce (1730–94) discovered Lake Tana in the east, in what is now Ethiopia. He realized that it was the source of the Blue Nile, one of the main tributaries of the great River Nile. In the west, Mungo Park (1771–1806) set out in 1795 to explore the mysterious, little-known River Niger, which flowed inland and never seemed to reach the sea. He discovered that the river actually flowed east, not west as had always been thought, and that it turned south near Timbuktu. However, he was not clear what happened to it after that. He drowned when his canoe was ambushed by tribesmen.

Over the next fifty years attention turned to the Sahara Desert. In 1828 a French explorer called René Caillié (1799–1838) became the first European to survive a secret visit to the legendary and forbidden city of Timbuktu, which as a Muslim city was closed to Christians. He was very

◀ TIMBUKTU
During the 1300s the city of Timbuktu became a prosperous center for trade across the Sahara Desert. Over the centuries, the city became famous for its wealth and learning, although no European had ever visited it.

THE SLAVE TRADE
The first black slaves were shipped out of Africa by the Arabs more than 1,000 years ago. Local rulers grew rich by selling captured enemies into slavery. In 1482 the Portuguese opened a trading post for exporting slaves to the New World. Other European nations joined in. From 1701 to 1810 more than seven million Africans were sent to the Americas. In the early 1800s slavery was abolished in Europe, but the Arabs continued the trade until 1873, when the main slave market in Zanzibar was closed.

▲ CATCHING SLAVES
In West Africa armed slave traders captured young African men and took them to the slave ports ready for export.

▼ LIFE OF A SLAVE
Slaves worked very long hours, six days a week on the plantations of the New World. If a slave tried to escape, they were made to wear a heavy iron collar with long spikes, which made it difficult for them, if they tried again, to escape to freedom through the undergrowth.

▼ SLAVE SHIPS
Slaves were packed into the holds of ships for transportation across the Atlantic. Conditions were bad and more than one million people died on the way.

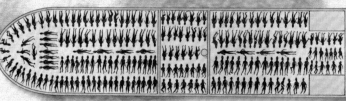

Deep Inside Africa

▶ INTO AFRICA
From the 1760s onwards, Europeans made determined attempts to explore the interior of Africa. The major rivers of the Niger and Nile were mapped and the great Sahara Desert was thoroughly explored. The south of the continent, however, remained largely unknown.

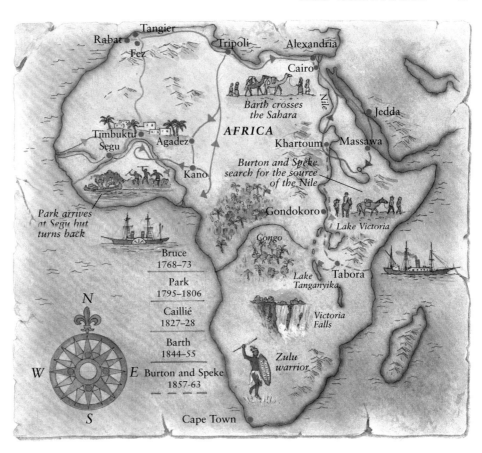

disappointed to find that it was full of mud huts, not rich buildings, and few people believed him when he returned to France. His story was confirmed, however, by the German explorer Heinrich Barth (1821–65), who explored the entire region thoroughly for the British government during the 1850s.

In 1857 two hardy British explorers, Richard Burton (1821–90) and his friend John Speke (1827–64), set out to solve one of the great African mysteries – the source of the River Nile. They explored the great lakes in East Africa, but Burton fell ill and Speke continued alone. After two attempts, he discovered that the Nile flowed out of the northern end of Lake Victoria (which Speke named after the reigning British queen) and over the mighty Ripon Falls. The interior of Africa was slowly giving up its secrets.

▲ TRANSPORT
The slave traders rarely travelled into the interior of Africa. Slaves would be brought to the coast for sale by tribal leaders who had enslaved their enemies. The slaves would either have been forced to walk to the coast, or taken in canoes such as this one.

▼ SAHARA DESERT
Even the vast Sahara Desert was not a barrier to slave traders who would take caravans of slaves across the desert, empty but for occasional rock formations like this one.

Key Dates

- 1768–73 James Bruce searches for the source of the Nile.
- 1795–1806 Mungo Park explores the River Niger.
- 1827–28 René Caillié becomes the first European to visit the city of Timbuktu.
- 1844–55 Heinrich Barth travels across the Sahara Desert.
- 1857–58 Richard Burton and John Speke explore the great lakes of East Africa.
- 1858–63 John Speke investigates the Nile and eventually discovers its source.

Livingstone and Stanley

▲ LION ATTACK
In 1844 Livingstone was attacked by a lion, which mauled his left shoulder. Although he eventually got better, he never regained the full use of his left arm.

ONE MAN MORE THAN any other transformed European knowledge about Africa – David Livingstone (1813–73). He started out as a missionary and doctor, and went to Africa to convert the local people to Christianity and to improve their lives through medicine and education. Once in Africa, however, Livingstone became curious about everything he saw and began to travel extensively. He recorded it all in three vast books, totalling more than 750,000 words, which made him and his journeys world-famous. But today some people think that his travels were not such a good thing, because he paved the way for the European colonization and exploitation of Africa.

Livingstone was born in Scotland and arrived in Cape Town on the Cape of Good Hope in 1841. From there he travelled to the mission station of Kuruman, on the edge of the Kalahari Desert. Here he met his future wife, Mary, and had a family. Together they established further missions, but Livingstone soon got restless and sent his wife and children back to England so that he could continue to explore by himself.

In 1851 he discovered the River Zambezi, which was previously unknown to people in Europe. From 1852 to '56 he became the first European to cross the continent from east to west, exploring the length of the Zambezi as he went. He then investigated the eastern coast and in 1865 set out to find the source of the River Nile. For a while, nothing was heard of him and an American newspaper, *The New York Herald*, sent a reporter, Welsh-born Henry Stanley (1841–1904), to find him. He managed to track him down eight months later. Stanley returned twice to Africa, once to explore the Great Lakes region and sail down the last great unknown river in

◀ RIVER SCARES
Livingstone made many of his journeys by boat, braving rapids, waterfalls, and even hippopotamuses. On one occasion, his boat crashed into a hippo and was overturned, causing him to lose much of his equipment.

THE INTERIOR OF AFRICA
John Speke, who discovered the source of the River Nile, described Africa as an upside-down soup plate – a rim of flat land round the edge with a sharp rise up to a central plateau. Rivers flowing from the interior often crashed over rapids or waterfalls, and explorers had to carry their boats round them. Wild animals and hostile locals added to the problems.

◀ VICTORIA FALLS
On November 17, 1855, Livingstone came up against a huge waterfall on the River Zambezi. Clouds of water vapour gave it its local name of Mosi-oa-tunya, or "the smoke that roars." Livingstone named the falls after Queen Victoria of Britain, the only English name he gave to any discovery.

▲ STANLEY'S HAT
Stanley wore a hat like this one, and Livingstone a flat cap, at their famous meeting at Ujiji.

▲ ZULU WOMEN
The Zulus of southern Afri were a warlike people. Und their leader Shaka, they bu a powerful nation in the region in the early 1800s. Today there are more than seven million Zulu people living in South Africa.

LIVINGSTONE AND STANLEY

◀ "DR LIVINGSTONE, I PRESUME?"
In March 1871 Henry Stanley set out from Zanzibar to find Dr Livingstone. Stanley was an adventurer who was probably motivated by fame and fortune. Eight months later he heard from local people that Livingstone had recently returned to Ujiji, on the shores of Lake Tanganyika. Stanley rushed to meet the ailing Livingstone, greeting him on November 10, with the now famous words: "Dr Livingstone, I presume?" "Yes," replied the explorer.

Africa, the River Congo (now Zaïre). Then, after working for the Belgian king in the Congo, he went to rescue Emin Pasha, the British governor of Equatoria (north of the Great Lakes), who was besieged by enemy tribespeople. His expeditions helped both Britain and Belgium establish colonies in central Africa, and by the time he died in 1904, almost all of Africa was under European control.

▶ DR LIVINGSTONE AND MR STANLEY
Between them Livingstone and Stanley explored much of central and southern Africa. They also navigated the two great and previously unknown African rivers – the Zambezi and Congo (now Zaïre) – although neither proved very easy to navigate. They also explored the Great Lakes region, confirming the source of the River Nile and settling disputes about how the different lakes drained into each other.

Key Dates

- 1841–52 Livingstone explores southern Africa.
- 1852–56 Livingstone goes down the River Zambezi. Discovers Victoria Falls.
- 1858–64 Livingstone explores Lake Nyasa and eastern coast.
- 1865–73 Livingstone goes to the Great Lakes and disappears.
- 1871–72 Stanley looks for Livingstone. Finds him in Ujiji.
- 1874–77 Stanley goes down the River Congo to Cabinda in west.
- 1887–89 Stanley rescues besieged Emin Pasha.

The North Pole

▲ ROBERT PEARY
The American explorer Robert Peary made eight journeys to the Arctic, finally reaching his goal, the North Pole, in 1909.

In 1881 a ship, the *Jeannette*, sank off the coast of Siberia. Three years later the wreckage turned up 2,983 miles (4,800 km) away on the coast of Greenland, right on the other side of the Arctic Ocean. This extraordinary event caused great confusion, because everybody knew that the Arctic Ocean consisted of a thick layer of pack ice. How had the wreckage managed to travel such a great distance? And how had it moved through the ice?

The Norwegian explorer Fridtjof Nansen (1861–1930) decided to find out. He calculated that the wreckage could only have been moved by a powerful ocean current which had pushed it along in the ice. Nansen designed a boat, the *Fram*, which he intended to steer into the ice and allow the currents to move it, just as they had the *Jeannette*. He worked out that the currents would carry him close to the North Pole, in the middle of the Arctic Ocean. For three years, the *Fram* drifted in the ice from Siberia to the Spitsbergen islands to the east of Greenland. Although he failed to reach the North Pole, Nansen did prove that there was no land under the North Pole – it was just ice.

Nansen was not the first explorer to try to reach the North Pole. Between 1861 and 1871 an American, Charles Hall (1821–71), made three attempts on foot, dying after his last journey. But it was Nansen's voyage that raised huge international interest in the North Pole and a race to get there first began.

In 1897 the Swedish engineer Salomon Andrée tried to fly to the North Pole in a balloon, but he perished soon after taking off from Spitsbergen. Robert Peary (1856–1920) was more successful. He was an American explorer who made his first visit to the Arctic in 1886. For the next 22 years he devoted himself to polar exploration, returning year after year, each time getting closer to his goal of reaching the North Pole. In

◀ HUSKIES
Husky dogs have a thick, double coat of fur which helps to keep them warm in the extreme cold and snowy conditions of the Arctic. They can be trained to pull sledges of equipment over the ice.

WHAT'S IT LIKE AT THE NORTH POLE?
The North Pole is situated in the middle of the Arctic Ocean, which is covered with pack ice all year round. Because the ice floats on top of the ocean currents it is broken up and jagged, often rising to ridges 33 feet (10 meters) or more high.

▲ REFUELLING AT THE NORTH POLE
Aircraft play a vital role in bringing in supplies to the North Pole. American explorers Richard Byrd and Floyd Bennett reached the North Pole by airplane in 1926.

▲ SEALSKIN
The first European travelers in the Arctic wore layers of woollen clothes, which failed to protect them from the cold. Later, they learnt to wear Inuit-style animal-skin clothes, such as this sealskin hood.

▼ PEMMICAN
An ideal food for a long Arctic expedition is pemmican. It is made from dried, shredded meat mixed with melted fat. It is full of calories and lasts for years.

THE NORTH POLE 53

Icebergs and frozen pack ice pile up either side of the *Fram*

◀ THE *FRAM*
Nansen needed a very strong ship for his plan. The Fram *was specially designed to be frozen into the Arctic ice so that it could then float with the currents across the Arctic Ocean without being damaged. Nansen hoped that the ice-bound ship would drift towards the North Pole. The ship went right across the Ocean, but it didn't get as near to the Pole as Nansen had hoped.*

Hull built to withstand the huge pressure of the ice

1908 he set off up the west coast of Greenland and established a base camp at Cape Columbia on Ellesmere Island. His six-strong team set off from there, making a mad dash and reaching the North Pole on April 6, 1909. They then hurried back to base camp. The top of the world had been conquered.

Some people doubted that Peary had reached the North Pole, since he made the return journey in record time, covering 70 miles (112 km) in one day. Nowadays most people think that Peary did indeed reach the Pole.

▼ USS *NAUTILUS*
In 1958 a US nuclear-powered submarine, the USS *Nautilus*, sailed under the polar ice-cap. It left Point Barrow in Alaska and sailed the 1,820 miles (2,930 km) to Spitsbergen in the North Atlantic Ocean in four days. The submarine, which was 299 feet (91 m) long and had a crew of 116, passed directly under the North Pole.

Key Dates

- 1871 Charles Hall sails up the west coast of Greenland and gets nearer to the North Pole than anyone before him.

- 1893–96 Fridtjof Nansen sails the *Fram* into the polar ice and drifts towards the North Pole, but fails to reach it.

- 1897 Salomon Andrée attempts to reach the Pole by balloon, but dies in the attempt.

- 1908–09 Robert Peary reaches the North Pole.

- 1958 USS *Nautilus* sails under the polar ice-cap.

Race to the South Pole

After Robert Peary made his successful attempt on the North Pole in 1909, all eyes turned towards the South Pole. Because it was the last unconquered place on Earth, the South Pole held a huge attraction for explorers, but it was a daunting place to visit.

Unlike the North Pole, the South Pole is covered by land. The vast, frozen continent of Antarctica is the coldest place on Earth, with ridges of mountains and large glaciers, making travelling extremely difficult. In addition, the land is surrounded by pack ice and icebergs that stretch far into the Southern Ocean. Two people prepared themselves to conquer this icy wilderness. The first was Robert Scott, a British explorer who had visited the region in 1901–4 and came to think of the continent as his to conquer. As his inten-

▶ ROALD AMUNDSEN
Norwegian polar adventurer Roald Amundsen (1872–1928) was a skilled explorer and had three impressive records to his name. He was the first person to sail through the Northwest Passage, in 1903–6, the first person to reach the South Pole, in 1911, and the first person to fly an airship across the North Pole, in 1926.

tion to lead an expedition to the South Pole became known, a second explorer, the Norwegian Roald Amundsen, joined the race. He kept his plans secret to prevent Scott speeding up his preparations. Amundsen too was an experienced polar explorer, but he was far better equipped and prepared than Scott.

Both expeditions arrived in Antarctica in January 1911 and spent the winter either side of the Ross Ice Shelf. Amundsen, however, had left two weeks before Scott and was 68 miles (110 km) closer to the Pole. He was also better prepared, having already made several journeys to leave food stores at stages along the route. His five-strong party made fast progress, climbing the steep Axel Heiberg glacier onto the plateau surround-

◀ PENGUINS
The Antarctic is home to several different species of penguin. Penguins cannot fly, but use their wings as flippers to swim.

SCOTT'S JOURNEY
In 1910 Robert Scott set out for Antarctica on board the ship *Terra Nova*. After spending the winter at Cape Evans, he set out for the South Pole in November 1911. Unlike his rival Amundsen, Scott used ponies as well as dogs to haul the sledges, but the ponies died in the cold. As a result, the party of five made slow progress and were devastated to discover, when they reached the South Pole on January 17, 1912, that Amundsen had beaten them to it. All five died on the return journey.

▲ ROBERT SCOTT
The naval officer Robert Scott (1869–1912) led a scientific expedition to Antarctica in 1901. His ill-fated expedition to the Pole in 1910–12 captured the imagination of the world.

▶ SCOTT'S BASE CAMP
Scott established his base camp at Cape Evans, on the east side of the Ross Ice Shelf. Here he and his team spent the winter of 1911, planning their route to the Pole, studying maps and also writing letters and reports.

◀ CHEMISTRY SET
Scott's expeditions were scientific as well as exploratory. His team carried this chemistry set with them when they set off in 1910.

Race to the South Pole 55

▶ BATTLING ACROSS A FROZEN LAND
Amundsen and his party were well equipped to endure the cold conditions and were all expert skiers. They used husky dogs to pull their sledges. As food and other provisions were used up and the sledges got lighter, unwanted dogs were shot and eaten, reducing the amount of food required for the expedition. As a result, Amundsen and his team travelled far faster than Scott's team.

ing the South Pole. They arrived at the Pole on December 14, 1911. Scott set out on November 1, 1911 but encountered far worse weather and made slow progress, finally getting to the Pole a month after Amundsen, on January 17, 1912.

Amundsen's expedition skills and equipment ensured that all his party returned home safely. Sadly, Scott and his team all perished, three of them within 11 miles (18 km) of a supply depot equipped with food and other life-saving provisions. The race to the South Pole was over, but although Amundsen claimed the prize, Scott has continued to hold a special fascination for people, because of the tragic ending to his expedition.

▼ RESEARCH STATION, ANTARCTICA
In 1959 an agreement was made to reserve Antarctica for scientific research. Today, 18 nations have scientific bases there to conduct research into the environment, wildlife and weather. In 1987 scientists found a hole in the ozone layer above Antarctica. The ozone layer protects the earth from the harmful rays of the sun.

Key Dates

- 1840 Antarctic coastline visited by James Wilkes and Jules Dumont d'Urville.
- 1841 James Ross from Britain explores the Ross Sea and its vast ice shelf.
- 1901–04 Scott explores the Antarctic coast and Ross Sea.
- 1908 Ernest Shackleton gets within 112 miles (180 km) of the Pole.
- 1911 Roald Amundsen reaches the South Pole.
- 1912 Scott gets to the Pole but the team dies on return journey.

Seas, Summits and Skies

▲ CHARLES LINDBERGH
The first solo flight across the Atlantic was made by 25-year-old Charles Lindbergh in 1927, when he flew the Spirit of St Louis *from New York to Paris in 33 hours.*

With the conquest of the South Pole in 1911, an age of exploration came to an end. All the major undiscovered parts of the world had now been explored. But eight years earlier, in 1903, a new method of transport had made its début. Orville and Wilbur Wright took to the skies over North Carolina in the airplane they had built, called the *Flyer*. Powered aircraft created new opportunities for exploration and discovery and in the first 30 years of the 20th century a series of epic flights took place. Louis Blériot made the first crossing of the English Channel in 1909. The first non-stop journey across the North Atlantic, from Newfoundland to Ireland, followed a decade later. It was made by John Alcock and Arthur Brown. Charles Lindbergh flew solo over the Atlantic in 1927, while Amy Johnson made the first solo flight by a woman, from Britain to Australia in 1930. These and other historic flights opened up the skies to commercial travel, and airlines began regular flights between the major cities of the world. People were now able to travel to faraway places without spending months at sea in order to get there.

▲ BALLOONING ROUND THE WORLD
In 1999 Brian James and Bertrand Piccard became the first people to circumnavigate the world non-stop in their balloon, the Breitling Orbiter 3. *The pair set off from Switzerland and used the jet streams in the upper atmosphere to glide eastwards around the world.*

As a result, more and more people decided to travel to other countries and explore the world for themselves. The growth in foreign travel led to a change in the nature of exploration. Now explorers took to the air, surveying lands by airplane and producing detailed maps by aerial reconnaissance. For those expeditions still on foot, supplies and reinforcements could now be

HIGHS AND LOWS
Although nearly three-quarters of the world's surface is covered by sea, we still know very little about what lies beneath the ocean's surface. The development of underwater craft enabled explorers to study the seas in greater detail. At the other extreme, the highest places on the Earth's surface have similarly fascinated explorers.

▲ THE *TRIESTE*
The *Trieste* bathyscaphe was designed to withstand the great pressure under the sea. In 1960 Jacques Piccard descended 7 miles (11km) into the Marianas Trench in the western Pacific Ocean, setting a world record that survives today.

▶ JACQUES COUSTEAU
One of the world's most famous ocean explorers, Jacques Cousteau (1910–97) invented the aqualung in 1943 to help divers breathe underwater. It was an air tank connected to a face mask.

▲ DIVING BELL
Edmund Halley invented the diving bell in 1690. It consisted of a watertight barrel anchored to the sea floor by heavy weights. Barrels of air were lowered and connected to the bell to supply the divers with fresh air.

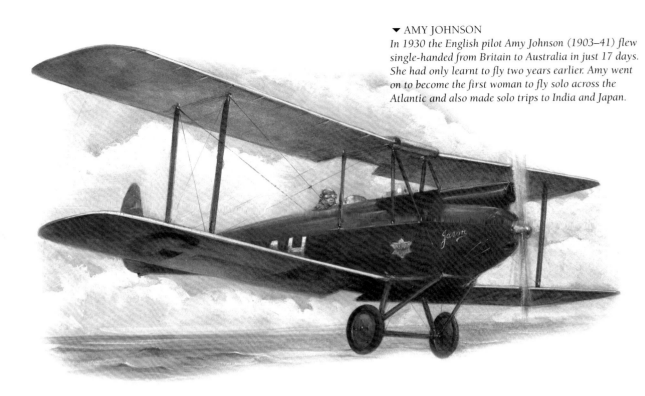

▼ AMY JOHNSON
In 1930 the English pilot Amy Johnson (1903–41) flew single-handed from Britain to Australia in just 17 days. She had only learnt to fly two years earlier. Amy went on to become the first woman to fly solo across the Atlantic and also made solo trips to India and Japan.

airlifted in, and any casualties flown out for medical treatment. As a result, explorers face less physical danger than they used to, and their emphasis has shifted away from exploration from its own sake towards exploration for scientific reasons. Today, teams of scientists investigate the impact of global warming in Antarctica, for example, or the effects of the climate change in the Pacific Ocean. They use highly complex scientific instruments and techniques, and have a support team ready to fly them out of danger at a moment's notice.

▶ MOUNT EVEREST
Climbing to the summit of the highest peak on Earth has always fascinated mountaineers. Mount Everest (29, 029 feet; 8848m) lies between Nepal and Tibet and mountaineers found it a very difficult challenge. Thirteen expeditions tried to reach the summit before the New Zealander Sir Edmund Hillary (born 1919) and the Nepalese Tenzing Norgay (1914–86) succeeded on May 29, 1953. Within a year, most of the other major Himalayan peaks were also conquered by European mountaineers.

Key Dates

- 1903 Wright brothers' flight.
- 1909 Louis Blériot flies non-stop across the English Channel.
- 1919 Alcock and Brown cross the Atlantic.
- 1927 Charles Lindbergh crosses the Atlantic.
- 1930 Amy Johnson flies single-handed from Britain to Australia.
- 1953 Everest conquered.
- 1954 Italians climb K2 – world's second-highest mountain.
- 1960 Jacques Piccard descends to record depths below the sea.

Blasting into Space

On October 4, 1957 an aluminum sphere no bigger than a large beachball was launched into space by the USSR. It measured 1.9 feet (58 cm) across and had four antennae trailing behind it. It orbited the Earth once every 96 minutes. This was *Sputnik I*, the world's first artificial satellite, and it began a period of intense space exploration and discovery that continues to this day.

Modern rocket technology had made it possible to travel out of the Earth's atmosphere and into space. Once there, it became easier to fly to the Moon and to examine our closest planet neighbors in the solar system. Scientists wanted to find out the answer to some of the oldest questions on Earth – is there life elsewhere in the universe? How and when were the Earth, and the universe itself, formed?

▲ LAUNCH SITE
A rocket needs huge power to lift it and its load off the launch pad. Once in space, the rocket is no longer needed and falls away, leaving the spacecraft or satellite to continue on its own.

▲ FLOATING IN SPACE
Astronauts are able to venture outside their spacecraft to do repairs or to help it dock with another craft. They must be tethered to their own craft to stop them drifting off into space.

▶ MOON LIVING
Astronauts lived in this lunar module when they landed on the Moon. When they were ready to leave, the module blasted off to rejoin the orbiting main spacecraft.

THE SPACE RACE
The former USSR launched the world's first satellite in 1957, beginning a space race with the USA that lasted until 1969. The Americans feared that the USSR would use space for military purposes, and wanted to prove that the USA was the world's leading superpower. The race ended when the USA landed a man on the Moon. Today the two countries cooperate on missions.

▲ DOG IN SPACE
The first living creature in space – a Russian dog called Laika – was launched into space on board *Sputnik 2* in November 1957 and remained in orbit for two days. Many other creatures, such as monkeys and jellyfish, have made the trip.

▶ SPACE FOOD
Pre-packed, specially prepared food is taken on space missions. It requires heat or water to make it edible. Fresh foods are rarely eaten because they do not keep well.

◀ YURI GAGARIN
The first human to go into space was the Russian cosmonaut Yuri Gagarin (1934–68). On April 12, 1961 he orbited the Earth once on board *Vostok 1*, returning to Earth after 108 minutes in space. Gagarin became a hero throughout the USSR and was given many national honors.

Blasting into Space   59

◀ MOON WALK
Neil Armstrong became the first person to walk on the Moon on July 24, 1969. He said, "That's one small step for man, one giant leap for mankind." Today, only 12 astronauts, including Buzz Aldrin, pictured, have been there.

▼ WORKING IN SPACE
The space shuttle is launched like a rocket, but returns to Earth like a plane. It can then be used again. In space the shuttle is used for launching, repairing and recovering satellites and for further scientific research.

They also wanted to explore the nearest planets and find out more about them.

This combination of technology and curiosity has sent men to the Moon and unmanned spacecraft to examine every planet in the solar system. Weather, communication and spy satellites now orbit the Earth in huge numbers. At least two new satellites are launched each week. Orbiting telescopes send back detailed information about distant stars, and permanent space stations enable astronauts to spend many months in space. Gradually a more complete picture is being built up about our solar system and its place in the universe, and new discoveries are made every year.

▼ THE HUBBLE TELESCOPE
In 1990 the *Hubble* space telescope was launched into orbit high above the Earth. It sends back X-ray and other photographs free from interference or distortion by the Earth's atmosphere.

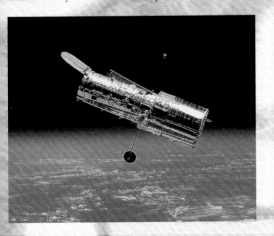

▲ THE GALAXY
The exploration of space has told scientists more about our own galaxy (the Milky Way) and the millions of stars it contains. By observing how these stars are born and die, scientists have begun to understand how the universe itself was formed.

Key Dates

- 1957 Russians launch *Sputnik I*, the first satellite, into space.
- 1960 First weather, navigation, communication satellites (USA).
- 1961 Soviet cosmonaut Yuri Gagarin is first person in space.
- 1966 *Luna IX* lands on Moon.
- 1969 Neil Armstrong is first person to walk on the Moon.
- 1970 USSR launches *Salyut I*, the world's first space station.
- 1981 *Columbia* space shuttle.
- 1983 *Pioneer 10* is first space probe to leave solar system.

Glossary

A
archipelago A large or chain group of islands.
astrolabe Navigation instrument used by sailors to measure the height of the Sun at noon, thus giving the ship's latitude.

B
barbarian A term used by the Romans to describe the nomadic invaders of their empire.

C
caravan A group of people, usually riding camels, who travel together across the desert. Arabs crossing the Sahara Desert and merchants journeying along the Silk Road all travelled in caravans.
civilization A settled society that has developed writing, organized religion, trade, great buildings and a form of government.
colony A region or country that is controlled by another country. People who settle in a colony are known as colonists.
commerce Trade, or buying and selling goods on a large scale, usually for profit.
continent One of the seven great landmasses of the world: Europe, Asia, Africa, North and South America, Australasia and Antarctica.
Crusade A military expedition launched from Christian Europe to attempt to recover the Holy Land from Muslim rule.
culture The art, literature, music and painting of a country or civilization. It also describes the way of life of a particular country.

D
dynasty A ruling family.

E
empire A large area made of different lands or countries, ruled by one government or leader.
Equator An imaginary line that circles the middle of the Earth at an equal distance between the North and South Poles.
Eurasia The huge mass of land formed by Europe and Asia.

G
glacier A slow-moving river of ice.

J
Jesuit A member of a Christian religious order formed in 1534 to undertake missionary work.

K
kingdom A country or region ruled over by a king or queen and where power passes down through the generations of the royal family.

L
latitude Imaginary horizontal lines that circle the Earth and are measured in degrees north or south of the Equator. Navigators calculate latitude to know how far north or south they are.
longitude Imaginary vertical lines that circle the Earth and are measured in degrees east or west of a line called the prime meridian, which runs through Greenwich in England. Navigators calculate longitude to know how far east or west they are.

M
Mesopotamia The fertile area of land between the Tigris and Euphrates rivers in the Middle East. The word literally means "the land between two rivers."
missionary A person who travels abroad to convert the local people to his or her own religion.

N
nomad A member of a group of people who move from place to place in search of food, water and land for grazing their animals.
North Pole The northernmost point of the Earth's axis of rotation, which lies in the Arctic Ocean.
northern hemisphere The half of the Earth that lies to the north of the Equator.

P
pack ice A large amount of ice floating on a sea or ocean that has become packed together in huge sheets.
pelt The skin of a furry animal, such as a sable, otter or mink, which is made into coats, hats, gloves and boots.
peninsula A strip of land that is almost an island, surrounded by water on three sides.
pilgrim A person who goes on a long journey, or pilgrimage, to a holy place.

plateau A high, flat area of land.
polar Referring to the North or South Pole.

Q

quadrant Navigation instrument, consisting of a quarter-circle marked in degrees. This was used by sailors to calculate the angle of the Sun and thus to work out the ship's latitude.

R

republic A country or state, such as ancient Rome, ruled by elected representatives of its people.

S

seal An engraved disc used to leave an impression on soft wax. Government documents are sealed to show that they are official.

Silk Road The ancient trading route between China and Europe. It was the route by which Chinese silk reached Europe.

South Pole The southernmost point of the Earth's axis of rotation, which lies in Antarctica.

southern hemisphere The half of the Earth that lies to the south of the Equator.

superpower A country possessing immense economic and military power, such as the USA or the former USSR.

T

trade The process of buying and selling goods.

tribe A group of people usually descended from the same family or sharing the same language and culture, with a recognized leader.

tsar The title of the hereditary emperor and ruler of Russia. Also sometimes written as czar.

For More Information

American Museum of Natural History
Central Park West at 79th Street
New York, NY, 10024-5192
(212) 769-5100
Web site: http://www.amnh.org
The museum's mission is to discover, interpret, and disseminate—through scientific research and education—knowledge about human cultures, the natural world, and the universe.

Explorers Club
46 East 70th Street
New York, NY 10021
(212) 628-8383
Web site: http://www.explorers.org/index.php
The Explorers Club is a multidisciplinary professional society dedicated to the advancement of field research and the ideal that it is vital to preserve the instinct to explore. Since its inception in 1904, the club has served as a meeting point and unifying force for explorers and scientists worldwide. Founded in New York City in 1904, the Explorers Club promotes the scientific exploration of land, sea, air, and space by supporting research and education in the physical, natural, and biological sciences. The club's members have been responsible for an illustrious series of famous firsts: first to the North Pole, first to the South Pole, and first to the summit of Mount Everest, first to the deepest point in the ocean, first to the surface of the Moon.

Library of Congress
Geography and Map Division
101 Independtence Avenue SE
Washington, DC 20540-4650
(202) 707-6277
Web site:
http://memory.loc.gov/ammem/gmdhtml/dsxphome.html
The Geography and Map Division of the Library of Congress is the largest and most comprehensive cartographic collection in the world, numbering over 5.2 million maps, including 80,000 atlases, 6,000 reference works, numerous globes and three-dimensional plastic relief models, and a large number of cartographic materials in other formats, including electronic. Many of these maps reflect the European Age of Discovery, dating from the late 15th century to the 17th century when Europeans were concerned primarily with determining the outline of the continents as they explored and mapped the coastal areas and the major waterways. Also included are 18th and 19th century maps documenting the exploration and mapping of the interior parts of the continents, reflecting the work of Lewis and Clark and subsequent government explorers and surveyors.

Mariners' Museum
100 Museum Drive
Newport News, VA 23606
(757) 596-2222
Web site: http://www.mariner.org
For over seventy-five years, the history of the ocean and its relationship with humankind has been told and displayed in one of the largest maritime museums in the world. The museum's collection totals approximately 35,000 artifacts documenting nearly 3,000 years of mankind's experiences on the seas and waterways of the world.

National Aeronautic and Space Administration (NASA) History Division
Headquarters Building
300 East Street SW
Mail Code CO72
Washington, DC 20546
(202) 358-0384
Web site: http://history.nasa.gov/index.html
Since its inception in 1958, NASA has accomplished many great scientific and technological feats in air and space. NASA technology also has been adapted for many non-aerospace uses by the private sector. NASA remains a leading force in scientific research and in stimulating public interest in aerospace exploration, as well as science and technology in general.

National Air and Space Museum
6th and Independence Avenue SW
Washington, DC 20560
(202) 633-1000
Web site: http://www.nasm.si.edu
The National Air and Space Museum on the National Mall in Washington, DC, has hundreds of original, historic artifacts on display, including the Wright 1903 flyer; the *Spirit of St. Louis*; the *Apollo 11* command module Columbia; and a lunar rock sample that visitors can touch. The museum offers twenty-two exhibition galleries, the Lockheed Martin IMAX Theater, flight simulators, a planetarium, tours, and daily free educational programs.

Web Sites
Due to the changing nature of Internet links, Rosen Publishing has developed an online list of Web sites related to the subject of this book. This site is updated regularly. Please use this link to access this list:

http://www.rosenlinks.com/jth/expl

For Further Reading

Adams, Simon. *The Kingfisher Atlas of Exploration and Empires.* New York, NY: Kingfisher, 2007.
Aronson, Marc, and John W. Glenn. *The World Made New: Why the Age of Exploration Happened and How It Changed the World.* Des Moines, IA: National Geographic Children's Books, 2007.
DK Publishing. *Atlas of Exploration.* New York, NY: DK Children, 2008.
Elliott, Lynn. *Exploration in the Renaissance.* New York, NY: Crabtree Publishing Co., 2009.
Fernández-Armesto, Felipe. *Pathfinders: A Global History of Exploration.* New York, NY: W. W. Norton & Co., 2007.
Fleming, Fergus. *Off the Map: Tales of Endurance and Exploration.* New York, NY: Grove Press, 2006.
Hannon, Sharon M. *Women Explorers.* Petaluma, CA: Pomegranate Communications, 2007.
Hernandez, Roger E. *Early Explorations: The 1500s.* Tarrytown, NY: Marshall Cavendish Children's Books, 2008.
Isserman, Maurice. *Across America: The Lewis and Clark Expedition.* New York, NY: Chlesea House Publications, 2009.
Oxford University Press. *Atlas of Exploration.* New York, NY: Oxford University Press, 2008.
Sansevere-Dreher, Diane. *Explorers Who Got Lost.* New York, NY: Tor Books, 2003.
Stott, Carole. *Space Exploration.* New York, NY: DK Children, 2009.

Index

A
Aborigines 44
Africa 18, 22–23, 48–51
aircraft 56–57
Al Idrisi 17
Alaska 33, 36, 37
Alexander the Great 8
Amazon 46–47
America 25
  Central 25, 26
  North 12, 30–31, 38–39, 58
  South 14, 25, 26–27, 28–29, 46–47
Amundsen, Roald 33, 54–55
Antarctica 43, 54–55
Arabs 16–17, 20, 48
Arctic Ocean 32, 34, 35, 43, 52, 53
Asia 8–9, 17, 18–19, 34–35, 36–37
Atlantic Ocean 24–25, 30
Attila 11
Australia 40–41, 42, 43, 44–45, 56
Aztecs 26–27, 29

B
balloons 52, 56
Barents, Willem 34, 35
Bering, Vitus 36–37
Bougainville, Louis 40, 41
Buddhism 11, 36, 37
Burton, Richard 48, 49

C
Cabeza de Vaca, Alvar Núñez 38, 39
Cabot, John 30
Caillié, René 48–49
Canada 30–31, 32–33, 38
Cartier, Jacques 30, 31
Chang Ch'ien 9
Charlemagne 11
China 8–9, 18–19, 36
Christianity
  conversion 5, 13, 36, 50
  Crusades 20, 21
  Dark Ages 10, 11
Clark, William 38–39
colonization
  Africa 50–51
  America 27, 31, 38
  Australia 44
  Phoenicians/Greeks 6, 7
  Polynesians 14
  Vikings 15
Columbus, Christopher 25–25
Compasses 23
Congo, River 51
Cook, Captain James 42–43
Cortés, Hernán 26
Crusades 20, 21

D
Da Gama, Vasco 22, 23
Dark Ages 10
De Balboa, Vasco 26, 27
de Champlain, Samuel 31
Dezhnev, Semyon 34, 35, 37
Drake, Francis 29

E
East Indies 40, 41
Easter Island 4
Egyptians 6, 7
Eric the Red 13
Eriksson, Leif 12
Europe 7, 8–9, 10, 10–11, 20–21
Everest, Mount 57
Eyre, Edward 44, 45

F
France, colonies 38
Franks 10, 11
Frobisher, Martin 32

G
Genghis Khan 18, 19
gold 26, 28, 29, 45
Great Lakes 31, 38, 39
Greeks 7
Greenland 12, 13, 32, 52, 53

H
Hanno 6, 7
Henry "The Navigator" 22, 23
Heyerdahl, Thor 14, 15
Hudson, Henry 32, 33
Humboldt, Alexander von 46
Huns 10, 11

I
Ibn Battuta 16, 18
Iceland 7, 12, 13
Incas 26, 27
India 9, 17, 23, 36
Indian Ocean 17, 19, 23, 40
Inuit people 32, 52
Islam 16–17

J
Japan 36, 37
Jesuits 36, 37, 38
Johnson, Amy 56, 57

K
Kelly, Ned 45
Kon-Tiki 16, 15
Kublai Khan 18, 19

L
La Salle, Robert de 38, 39
Lewis, Meriwether 38–39
Lindbergh, Charles 56
Lindisfarne Priory 10, 13
Livingstone, David 50–51
Louisiana 38

M
Machu Picchu 27
Magellan, Ferdinand 28, 29
Maoris 42
Mediterranean 6, 10, 13
Mexico 26, 27, 29, 38
Middle East 6, 13, 16, 17, 20
Mississippi, River 38, 39
Mohammed, the Prophet 16, 17
Mongols 18–19
Muslims *see* Islam

N
Nansen, Fridtjof 52, 53
Native Americans 30, 38
navigation 5, 12, 14, 17, 22–23

New World 25, 26–27
New Zealand 14, 40, 41, 42, 43
Niger, River 48
Nile, River 6, 7, 48, 49, 50
Nordenskjöld, Nils 34, 35
North Pole 52–53
Northeast Passage 34–36
Northwest Passage 32–33
Norway 7, 12, 13

O
ocean exploration 56
O'Hara Burke, Robert 44–45
Ostrogoths 10, 11

P
Pacific Ocean 14–15, 29, 34, 35, 36, 40–43
Peary, Robert 52, 53
Persian Empire 8
Peru 16, 26–27
Peter I, Tsar of Russia 36
Phoenicians 6–7
pilgrims 16, 17, 20, 48
pirate ships 28
Pizarro, Francisco 26, 27
Polo, Marco 19
Polynesians 14–15
Portugal 10, 22–23, 28, 48
Pytheas 7

Q
Quebec 31

R
rainforest 46–47
Romans 8, 10–11
Russia 13, 34–35, 36–37, 58

S
Sahara Desert 48, 49
St Lawrence River 30, 38
scientific discovery 4, 5
 aircraft impact 57
 Amazon 46–48
 Antarctica 54, 55
 Arabs 17
 Bering's expedition 36
 Captain Cook 43
 space travel 59
Scott, Robert 54, 55
ships 4–5
 caravel 22–23
 Christopher Columbus 25
 dhows 17
 Egyptians 6, 7
 first circumnavigation 28–29
 junks 18
 North Pole 53
 Northeast Passage 34–35
 Phoenicians 6–7
 Polynesians 14–15
 Vikings 12–13
Siberia 34–35, 36, 37
Silk Road 8–9, 18
slave trade 48–49
South Pole 42, 54–55
space exploration 5, 58–59
Spain 10, 17, 24–25, 28, 38
Speke, John 49
Stanley, Henry 50–51
Stuart, John 45
Sturt, Charles 44, 45
Sweden, Vikings 12–13

T
Tasman, Abel 40, 41
Tibet 36, 37, 57
Timbuktu 48–49
Torres, Luis 41
trade 4–5, 6, 8–9, 12–13, 16–17, 20–23

U
United States of America 38–39, 58

V
Vandals 10, 11
Vespucci, Amerigo 25
Vikings 12–13
Visigoths 10, 11

W
West Indies 24, 25
Wills, William 44–45

X
Xavier, Francis 36, 37

Z
Zambezi, River 50, 51
Zheng He 18, 19

About the Author

Simon Adams is a writer and editor of nonfiction books and has written extensively about history for children. His published works include *Castles and Forts* and *Archaeology Detectives*.